Unstructured Data Analysis
Entity Resolution and Regular Expressions in SAS®

K. Matthew Windham

sas.com/books

Contents

About This Book

What Does This Book Cover?

This book was written to provide readers with an introduction to the vast world that is unstructured data analysis. I wanted to ensure that SAS programmers of many different levels could approach the subject matter here, and come away with a robust set of tools to enable sophisticated analysis in the future.

I focus on the regular expression functionality that is available in SAS, and on presenting some basic data manipulation tools with the capabilities that SAS has to offer. I also spend significant time developing capabilities the reader can apply to the subject of entity resolution from end to end.

This book does not cover enterprise tools available from SAS that make some of the topics discussed herein much easier to use or more efficient. The goal here is to educate programmers, and help them understand the methods available to tackle these things for problems of reasonable scale. And for this reason, I don't tackle things like entity resolution in a "big data" context. It's just too much to do in one book, and that would not be a good place for a beginner or intermediate programmer to start.

Performing an array of unstructured data analysis techniques, culminating in the development of an entity resolution analytics framework with SAS code, is the central focus of this book. Therefore, I have generally arranged the chapters around that process. There is foundational information that must be covered in order to enable some of the later activities. So, Chapters 1 and 2 provide information that is critical for Chapter 3, and that is very useful for later chapters.

Chapter 1: Getting Started with Regular Expressions

In order to effectively prepare you for doing advanced unstructured data analysis, you need the fundamental tools to tackle that with SAS code. So, in this chapter, I introduce regular expressions.

Chapter 2: Using Regular Expressions in SAS

In this chapter, I will begin using regular expressions via SAS code by introducing the SAS functions and call routines that allow us to accomplish fairly sophisticated tasks. And I wrap up the chapter with some practical examples that should help you tackle real-world unstructured data analysis problems.

Chapter 3: Entity Resolution Analytics

I will introduce entity resolution analytics as a framework for applying what was learned in chapters 1 and 2 in combination with techniques introduced in the subsequent chapters of this book. This framework will be guiding force through the remaining chapters of this book, providing you with an approach to begin tackling entity resolution in your environment.

Chapter 4: Entity Extraction

Leveraging the foundation established in Chapters 1 and 2, I will discuss methods for extracting entity references from unstructured data sources. This should be a natural extension of the work that was done in Chapter 2, with a particular focus—preparing for the entity resolution.

Chapter 5: Extract, Transform, Load

I will cover some key ETL elements needed for effective data preparation of entity references, and demonstrate how they can be used with SAS code.

Chapter 6: Entity Resolution

In this chapter, I will walk you through the process of actually resolving entities, and acquaint you with some of the challenges of that process. I will again have examples in SAS code.

Chapter 7: Entity Network Mapping and Analysis

This chapter is focused on the steps taken to construct entity networks and analyze them. After the entity networks have been defined, I will walk through a variety of analyses that might be performed at this point (this is not an exhaustive list).

Chapter 8: Entity Management

In this chapter, I will discuss the challenges and best practices for managing entities effectively. I try to keep these guidelines general enough to fit within whatever management process your organization uses.

Appendix A: Additional Resources

I have included a few sections for random entity generation, regular expression references, Perl version notes, and binary/hexadecimal/ASCII code cross-references. I hope they prove useful references even after you have mastered the material.

Is This Book for You?

I wrote this book for ambitious SAS programmers who have practical problems to solve in their day-to-day tasks. I hope that it provides enough introductory information to get you started, motivational examples to keep you excited about these topics, and sufficient reference material to keep you referring back to it.

To make the best use of this book, you should have a solid understanding of Base SAS programming principles like the DATA step. While it is not required, exposure to PROC SQL and macros will be helpful in following some of the later code examples.

This book has been created with a fairly wide audience in mind—students, new SAS programmers, experienced analytics professionals, and expert data scientists. Therefore, I have provided information about both the business and technical aspects of performing unstructured data analysis throughout the book. Even if you are not a very experienced analytics professional, I expect you will gain an understanding of the business process and implications of unstructured data analysis techniques.

At a minimum, I want everyone reading this book to walk away with the following:

- A sound understanding of what both regular expressions and entity resolution are (and aren't)
- An appreciation for the real-world challenges involved in executing complex unstructured data analysis
- The ability to implement (or manage an implementation) of the entity resolution analytics methodology discussed later in this book

●   An understanding of how to leverage SAS software to perform unstructured data analysis for their desired applications

The SAS Platform is quite broad in scope and therefore provides professionals and organizations many different ways to execute the techniques that we will cover in this book. As such, I can't hope to cover every conceivable path or platform configuration to meet an organization's needs. Each situation is just different enough that the SAS software required to meet that organization's scale, user skill level(s), financial parameters, and business goals will vary greatly.

Therefore, I am presenting an approach to the subject matter which enables individuals and organizations to get started with the unstructured data analysis topics of regular expressions and entity resolution. The code and concepts developed in this book can be applied with solutions such as SAS Viya to yield an incredible level of flexibility and scale. But I am limiting the goals to those that can yield achievable results on a small scale in order for the process and techniques to be well understood. Also, the process for implementation is general enough to be applied to virtually any scale of project. And it is my sincere hope that this book provides you with the foundational knowledge to pursue unstructured data analysis projects well beyond my humble aim

What Should You Know about the Examples?

This book includes tutorials for you to follow to gain hands-on experience with SAS.

Software Used to Develop the Book's Content

SAS Studio (the same programming environment as SAS University Edition) was used to write and test all the code shown in this book. The functions and call routines demonstrated are from Base SAS, SAS/STAT, SAS/GRAPH, and SAS/OR.

Example Code and Data

You can access the example code and data for this book from the author page at https://support.sas.com/authors. Look for the cover thumbnail of this book and select "Example Code and Data."

SAS University Edition

If you are using SAS University Edition to access data and run your programs, check the SAS University Edition page to ensure that the software contains the product or products that you need to run the code: www.sas.com/universityedition.

At the time of printing, everything in the book, with the exception of the code in chapter 7, can be run with SAS University Edition. The analysis performed in chapter 7 uses procedures that are available only through SAS/OR.

About the Author

 Matthew Windham is a Principal Analytical Consultant in the SAS U.S. Government and Education practice, with a focus on Federal Law Enforcement and National Security programs. Before joining SAS, Matthew led teams providing mission-support across numerous federal agencies within the U.S. Departments of Defense, Treasury, and Homeland Security. Matthew is passionate about helping clients improve their daily operations through the application of mathematical and statistical modeling, data and text mining, and optimization. A longtime SAS user, Matthew enjoys leveraging the breadth of the SAS Platform to create innovative analytics solutions that have operational impact. Matthew is a Certified Analytics Professional, received his BS in Applied Mathematics from NC State University, and received his MS in Mathematics and Statistics from Georgetown University.

Learn more about this author by visiting his author page at https://support.sas.com/en/books/authors/matthew-windham.html. There you can download free book excerpts, access example code and data, read the latest reviews, get updates, and more.

We Want to Hear from You

SAS Press books are written *by* SAS Users *for* SAS Users. We welcome your participation in their development and your feedback on SAS Press books that you are using. Please visit sas.com/books to do the following:

- Sign up to review a book
- Recommend a topic
- Request information on how to become a SAS Press author
- Provide feedback on a book

Do you have questions about a SAS Press book that you are reading? Contact the author through saspress@sas.com or https://support.sas.com/author_feedback.

SAS has many resources to help you find answers and expand your knowledge. If you need additional help, see our list of resources: sas.com/books.

Acknowledgments

To my brilliant wife, Lori, thank you for always supporting and encouraging me in everything that I do. Thank you also to Bonnie and Thomas for always brightening my day. To my friends and family, your advice and encouragement have been treasured.

And I would like to thank the entire editorial team at SAS Press. Your collective patience, insight, and hard work have made this another wonderful writing experience.

Chapter 1: Getting Started with Regular Expressions

1.1 Introduction

This chapter focuses entirely on developing your understanding of regular expressions (RegEx) before getting into the details of using them in SAS. We will begin actually implementing RegEx with SAS in Chapter 2. It is a natural inclination to jump right into the SAS code behind all of this. However, RegEx patterns are fundamental to making the SAS coding elements useful. Without my explaining RegEx first, I could discuss the forthcoming SAS functions and calls only at a very theoretical level, and that is the opposite of what I am trying to accomplish. Also, trying to learn too many different elements of any process at the same time can simply be overwhelming for you.

To facilitate the mission of this book—practical application—without overwhelming you with too much information at one time (new functions, calls, and expressions), I will present a short bit of test code to use with the RegEx examples throughout the chapter. I want to stress the point that obtaining a thorough understanding of RegEx syntax is critical for harnessing the full power of this incredible capability in SAS.

1.1.1 Defining Regular Expressions

Before going any further, we need to define *regular expressions*.

Taking the very formal definition might not provide the desired level of clarity:

Definition 1 (formal)
   regular expressions: "Regular expressions consist of constants and operator symbols that denote sets of strings and operations over these sets, respectively."[1]

In the pursuit of clarity, we will operate with a slightly looser definition for regular expressions. Since practical application is our primary aim, it doesn't make sense to adhere to an overly esoteric definition. So, for our purposes we will use the following:

Definition 2 (informal, easier to understand)
   regular expressions: character patterns used for automated searching and matching.

In SAS programming, regular expressions are seen as strings of letters and special characters that are recognized by certain built-in SAS functions for the purpose of searching and matching. Combined with other built-in SAS functions and procedures, you can realize tremendous capabilities, some of which we explore in the next section.

Note: SAS uses the same syntax for regular expressions as the Perl programming language.[2] Thus, throughout SAS documentation, you find regular expressions repeatedly referred to as "Perl regular expressions." In this book, I chose the conventions that the SAS documentation uses, unless the Perl conventions are the most common to programmers. To learn more about how SAS views Perl, see the SAS documentation online.[3] To learn more about Perl programming, see the Perl programming documentation.[4] In this book, however, I primarily dispense with the references to Perl, as they can be confusing.

1.1.2 Motivational Examples

The information in this book is very useful for a wide array of applications. However, that will not become obvious until after you read it. So, in order to visualize how you can use this information in your work, I present some realistic examples.

As you are probably familiar with, data is rarely provided to analysts in a form that is immediately useful. It is frequently necessary to clean, transform, and enhance source data before it can be used—especially textual data. The following examples are devoid of the coding details that are discussed later in the book, but they do demonstrate these concepts at varying levels of sophistication. The primary goal here is to simply help you to see the utility for this information, and to begin thinking about ways to leverage it.

Extract, Transform, and Load (ETL)

ETL is a general set of processes for extracting data from its source, modifying it to fit your end needs, and loading it into a target location that enables you to best use it (e.g., database, data store, data warehouse). We're going to begin with a fairly basic example to get us started. Suppose we already have a SAS data set of customer addresses that contains some data quality issues. The method of recording the data is unknown to us, but visual inspection has revealed numerous occurrences of duplicative records, as in the table below. In this example, it is clearly the same individual with slightly different representations of the address and encoding for gender. But how do we fix such problems automatically for all of the records?

First Name	Last Name	DOB	Gender	Street	City	State	Zip
Robert	Smith	2/5/1967	M	123 Fourth Street	Fairfax,	VA	22030
Robert	Smith	2/5/1967	Male	123 Fourth St.	Fairfax	va	22030

Using regular expressions, we can algorithmically standardize abbreviations, remove punctuation, and do much more to ensure that each record is directly comparable. In this case, regular expressions enable us to perform more effective record keeping, which ultimately impacts downstream analysis and reporting.

We can easily leverage regular expressions to ensure that each record adheres to institutional standards. We can make each occurrence of Gender either "M/F" or "Male/Female," make every instance of the Street variable use "Street" or "St." in the address line, make each City variable include or exclude the comma, and abbreviate State as either all caps or all lowercase.

This example is quite simple, but it reveals the power of applying some basic data standardization techniques to data sets. By enforcing these standards across the entire data set, we are then able to properly identify duplicative references within the data set. In addition to making our analysis and reporting less error-prone, we can reduce data storage space and duplicative business activities associated with each record (for example, fewer customer catalogs will be mailed out, thus saving money!). For a detailed example involving ETL and how to solve this common problem of data standardization, see Section 2.4.1 in Chapter 2.

Data Manipulation

Suppose you have been given the task of creating a report on all Securities and Exchange Commission (SEC) administrative proceedings for the past ten years. However, the source data is just a bunch of .xml (XML) files, as shown in Figure 1.1. To the untrained eye, this looks like a lot of gibberish; to the trained eye, it looks like a lot of work.

Figure 1.1: Sample of 2009 SEC Administrative Proceedings XML File[5]

```
<?xml version="1.0" encoding="ISO-8859-1"?>
- <root>
    - <administrative_proceeding>
          <url>http://www.sec.gov/litigation/admin/2009/34-61262.pdf</url>
          <release_number>34-61262</release_number>
          <release_date>Dec. 30, 2009</release_date>
          <respondents>Stephen C. Gingrich</respondents>
      </administrative_proceeding>
    - <administrative_proceeding>
          <url>http://www.sec.gov/litigation/admin/2009/34-61256.pdf</url>
          <release_number>34-61256</release_number>
          <release_date>Dec. 30, 2009</release_date>
          <respondents>Gabelli Funds LLC</respondents>
      </administrative_proceeding>
    - <administrative_proceeding>
          <url>http://www.sec.gov/litigation/admin/2009/34-61255.pdf</url>
          <release_number>34-61255</release_number>
          <release_date>Dec. 30, 2009</release_date>
          <respondents>Gabelli Funds LLC</respondents>
      </administrative_proceeding>
```

However, with the proper use of regular expressions, creating this report becomes a fairly straightforward task. Regular expressions provide a method for us to algorithmically recognize patterns in the XML file, parse the data inside each tag, and generate a data set with the correct data columns. The resulting data set would contain a row for every record, structured similarly to this data set (for files with this transactional structure):

Example Data Set Structure

Release_Number	Release_Date	Respondents	URL
34-61262	Dec 30, 2009	Stephen C. Gingrich	http://www.sec.gov/litigation/admin/2009/34-61262.pdf
...	...	...	...

Note: Regular expressions cannot be used in isolation for this task due to the potential complexity of XML files. Sound logic and other Base SAS functions are required in order to process XML files in general. However, the point here is that regular expressions help us overcome some otherwise significant challenges to processing the data. If you are unfamiliar with XML or other tag-based languages (e.g., HTML), further reading on the topic is recommended. Though you don't need to know them at a deep level in order to process them effectively, it will save a lot of heartache to have an appreciation for how they are structured. I use some tag-based languages as part of the advanced examples in this book because they are so prevalent in practice.

Data Enrichment

Data enrichment is the process of using the data that we have to collect additional details or information from other sources about our subject matter, thus enriching the value of that data. In addition to parsing and structuring text, we can leverage the power of regular expressions in SAS to enrich data.

So, suppose we are going to do some economic impact analysis of the main SAS campus—located in Cary, NC—on the surrounding communities. In order to do this properly, we need to perform statistical analysis using geospatial information.

The address information is easily acquired from www.sas.com. However, it is useful, if not necessary, to include additional geo-location information such as latitude and longitude for effective analysis and reporting of geospatial statistics. The process of automating this is non-trivial, containing advanced programming steps that are beyond the scope of this book. However, it is important for you to understand that the techniques described in this book lead to just such sophisticated capabilities in the future. To make these techniques more tangible, we will walk through the steps and their results.

1. Start by extracting the address information embedded in Figure 1.2, just as in the data manipulation example, with regular expressions.

Figure 1.2: HTML Address Information

```
<p>World Headquarters<br>
SAS Institute Inc.<br>
100 SAS Campus Drive<br>
Cary, NC 27513-2414, USA<br>
Phone:919-677-8000<br>
Fax:919-677-4444<br>
</p>
```

Example Data Set Structure

Location	Address Line 1	Address Line 2	City	State	Zip	Phone	Fax
World Headquarters	SAS Institute Inc.	100 SAS Campus Drive	Cary	NC	27513-2414	919-677-8000	919-677-4444

2. Submit the address for geocoding via a web service like Google or Yahoo for free processing of the address into latitude and longitude. Type the following string into your browser to obtain the XML output, which is also sampled in Figure 1.3.

    http://maps.googleapis.com/maps/api/geocode/xml?address=100+SAS+Campus+Drive,+Cary,+NC&sensor=false

Figure 1.3: XML Geocoding Results

```
- <geometry>
    - <location>
        <lat>35.8301733</lat>
        <lng>-78.7664916</lng>
    </location>
    <location_type>ROOFTOP</location_type>
    - <viewport>
        - <southwest>
            <lat>35.8288243</lat>
            <lng>-78.7678406</lng>
        </southwest>
        - <northeast>
            <lat>35.8315223</lat>
            <lng>-78.7651426</lng>
        </northeast>
    </viewport>
</geometry>
```

3. Use regular expressions to parse the returned XML files for the desired information—latitude and longitude in our case—and add them to the data set.

   Note: We are skipping some of the details as to how our particular set of latitude and longitude points are parsed. The tools needed to perform such work are covered later in the book. This example is provided here primarily to spark your imagination about what is possible with regular expressions.

   Example Data Set Structure

Location	...	Latitude	Longitude
World Headquarters	...	35.8301733	-78.7664916

4. Verify your results by performing a reverse lookup of the latitude/longitude pair that we parsed out of the results file using https://maps.google.com/. As you can see in Figure 1.4, the expected result was achieved (SAS Campus Main Entrance in Cary, NC).

Figure 1.4: SAS Campus Using Google Maps

Now that we have an enriched data set that includes latitude and longitude, we can take the next steps for out the economic impact analysis.

Hopefully, the preceding examples have proven motivating, and you are now ready to discover the power of regular expressions with SAS. And remember, the last example was quite advanced—some sophisticated SAS programming capabilities were needed to achieve the result end-to-end. However, the majority of the work leveraged regular expressions.

1.1.3 RegEx Essentials

RegEx consist of letters, numbers, metacharacters, and special characters, which form patterns. In order for SAS to properly interpret these patterns, all RegEx values must be encapsulated by delimiter pairs—forward slash, /, is used throughout the text. (Refer to the test code in the next section). They act as the container for our patterns. So, all RegEx patterns that we create will look something like this: /pattern/.

For example, suppose we want to match the string of characters "Street" in an address. The pattern would look like /Street/. But we are clearly interested in doing more with RegEx than just searching for strings. So, the remainder of this chapter explores the various RegEx elements that we can insert into / / to develop rich capabilities.

Metacharacter

Before going any farther, some upcoming terminology should be clarified. *Metacharacter* is a term used quite frequently in this book, so it is important that it is clear what it actually means. A metacharacter is a character or set of characters used by a programming language like SAS for something other than its literal meaning. For example, \s represents a whitespace character in RegEx patterns, rather than just being a \ and the letter "s" that is collocated in the text. We begin our discussion of specific metacharacters in Section 1.3.

All nonliteral RegEx elements are some kind of metacharacter. It is good to keep this distinction clear, as I also make references to *character* when I want to discuss the actual string values or the results of metacharacter use.

Special Character

A *special character* is one of a limited set of ASCII characters that affects the structure and behavior of RegEx patterns. For example, opening and closing parentheses, (and), are used to create logical groups of characters or metacharacters in RegEx patterns. These are discussed thoroughly in Section 1.2.

RegEx Pattern Processing

At this juncture, it is also important to clarify how RegEx are processed by SAS. SAS reads each pattern from left to right in sequential *chunks*, matching each element (character or metacharacter) of the pattern in succession. If we want to match the string "hello", SAS searches until the first match of the letter "h" is found. Then, SAS determines whether the letter "e" immediately follows, and so on, until the entire string is found. Below is some pseudo code for this process, for which the logic is true even after we begin replacing characters with metacharacters (it would simply look more impressive).

Pseudo Code for Pattern Matching Process

```
START     IF POS = "h" THEN POS+1 NEXT ELSE POS+1 GOTO START
IF POS = "e" THEN POS+1 NEXT ELSE POS+1 GOTO START
   IF POS = "l" THEN POS+1 NEXT ELSE POS+1 GOTO START
   IF POS = "l" THEN POS+1 NEXT ELSE POS+1 GOTO START
   IF POS = "o" THEN MATCH=TRUE GOTO END ELSE POS+1 GOTO START
END
```

In this pseudo code, we see the START tag is our initiation of the algorithm, and the END tag denotes the termination of the algorithm. Meanwhile, the NEXT tag tells us when to skip to the next line of pseudo code, and the GOTO tag tells us to jump to a specified line in the pseudo code. The POS tag denotes the character position. We also have the usual IF, THEN, and ELSE logical tags in the code.

Again, this example demonstrates the search for "hello" in some text source. The algorithm initiates by testing whether the first character position is an "h". If it is not true, then the algorithm increments the character position by one—and tests for "h" again. If the first position is an "h", the character position is incremented, and the code tests for the letter "e". This continues until the word "hello" is found.

1.1.4 RegEx Test Code

The following code snippet enables you to quickly test new RegEx concepts as we go through the chapter. As you learn new RegEx metacharacters, options, and so on, you can edit this code in an effort to test the functionality. Also, more interesting data can be introduced by editing the `datalines` portion of the code. However, because we haven't yet discussed the details of how the pieces work, I discourage making edits outside the marked places in the code in order to avoid unforeseen errors arising at run time.

To keep things simple, we are using the DATALINES statement to define our data source and print the source string and the matched portion to the log. This should make it easier to follow what each new metacharacter is doing as we go through the text. Notice that everything is contained in a single DATA step, which does not generate a resulting data set (we are using _NULL_). The first line of our code is an IF statement that tests for the first record of our data set. The RegEx pattern is created only if we have encountered the first record in the data set, and is retained using the RETAIN statement. Afterward, the pattern reference identifier is reused by our code due to the RETAIN statement. Next, we pull in the data lines using the INPUT statement that assumes 50-character strings. Don't worry about the details of the CALL routine on the next line for now. We start writing SAS code in Chapter 2.

Essentially, the CALL routine inside the `RegEx Testing Framework` code shown below uses the RegEx pattern to find only the first matching occurrence of our pattern on each line of the `datalines` data. Finally, we use another IF statement to determine whether we found a pattern match. If we did, the code prints the results to the SAS log.

```
/*RegEx Testing Framework*/
data _NULL_;
if _N_=1 then
do;
    retain pattern_ID;
    pattern="/METACHARACTERS AND CHARACTERS GO HERE/"; /*<--Edit the pattern here.*/
    pattern_ID=prxparse(pattern);
end;
input some_data $50.;
call prxsubstr(pattern_ID, some_data, position, length);
if position ^= 0 then
    do;
        match=substr(some_data, position, length);
        put match:$QUOTE. "found in " some_data:$QUOTE.;
    end;
datalines;
Smith, BOB A.
ROBERT Allen Smith
Smithe, Cindy
103 Pennsylvania Ave. NW, Washington, DC 20216
508 First St. NW, Washington, DC 20001
650 1st St NE, Washington, DC 20002
3000 K Street NW, Washington, DC 20007
1560 Wilson Blvd, Arlington, VA 22209
1-800-123-4567
1(800) 789-1234
```

```
;
run;
```

Note: I have provided a jumble of data in the `datalines` portion of the code above. However, feel free to edit the data lines to thoroughly test each metacharacter as we go through this chapter.

Output 1.1 shows an example of the SAS log output provided by the previous code. For this example, I used merely the character string /Street/ for the pattern in order to create the output.

Output 1.1: Example Output Where pattern=/Street/

```
"Street" found in "3000 K Street NW, Washington, DC 20007"
```

The remaining information in this chapter provides a solid foundation for building robust, complex patterns in the future. Each element discussed is an independently useful building block for sophisticated text manipulation and analysis capabilities. Once we begin to combine these basic elements, we will create some very powerful analytic tools.

1.2 Special Characters

In addition to / (the forward slash), the characters ( ) | and \ (the backslash) are special and are thus treated differently than the RegEx metacharacters to be discussed later. Since some of these special characters are so fundamental to the structure of the RegEx pattern construction, we need to briefly discuss them first.

( )

The two parentheses create logical groups of pattern characters and metacharacters—the same way they work in SAS code for logic operations. It is important to create logical groupings in order to construct more sophisticated patterns. Nesting the parentheses is also possible.

|

The vertical bar represents a logical OR (much like in SAS). Again, the proper use of this element creates more sophisticated patterns. We will explore some interesting ways to use this character, starting with the example in Table 1.1. It is important to remember that the first item in an OR condition always matches before moving to the next condition.

\

The backslash is a tricky one as it has a couple of uses. It is used as an integral component of many other metacharacters (examples abound in Section 1.3). Think about it as an initiator that tells SAS, "Hey, this is a metacharacter, not just some letter." But that's not all it does. Since the special characters defined above also appear in text that we might want to process, the backslash also acts as a blocker that tells SAS, "Hey, treat this special character as just a regular character." By using \, we can create patterns that include parentheses, vertical bars, backslashes, forward slashes, and more—we simply add a \ in front of each occurrence of all the special characters that we want to treat as characters. For example, if we want our pattern to include open and closed parentheses respectively, the pattern would contain \(\).

Since you haven't learned any RegEx metacharacters yet, let's revisit strings using some of these new concepts. Notice that we can already start to match useful patterns with the characters and special characters.

**Table 1.1: Examples Using (), |, and **

Usage	Matches
/(C|c)at/	"Cat" "cat"
/cat|mouse/	"cat" "mouse"
/((S|s)treet)|((R|r)oad)/	"Street" "street" "Road" "road"
/\(This\)|\(That\) /	"(This)" or "(That)"

Note: In Perl parlance, \ is known as an *escape character*. To avoid any unnecessary confusion, we will dispense with this lingo and just refer to it as the backslash. However, be prepared to see that term used quite a bit in the Perl literature and on community websites.

Now, there are some additional special characters that also need the backslash in front of them in order to be matched as normal characters. They are: { } [ ] ^ $. * + and ?. All these characters are reserved and are thus treated differently, because they each have a special purpose and meaning in the world of RegEx. Since each one is defined and discussed at length in Sections 1.4 and 1.5, we will not discuss them further here. For now, just remember that they can't be used as part of pattern strings without the backslash immediately preceding them. Table 1.2 shows a few examples of how to use them as normal characters.

Table 1.2: Examples Using { } [ ] ^ $. * +

Usage	Matches
/\$1\.00 \+ \$0\.50 = \$1\.50/	"$1.00 + $0.50 = $1.50"
/2*3 = 6/	"2*3 = 6"
/\[2\]\^2/	"[2]^2"
/\{1,2,3,4,5\}/	"{1,2,3,4,5}"

Note: Notice that = and , match as characters (i.e., without a backslash) because they are not considered special characters.

1.3 Basic Metacharacters

As you write RegEx patterns in the future, you will find yourself using most of the metacharacters discussed in this section frequently because they are fundamental elements of RegEx pattern creation. Now, we can already build some useful patterns with the information discussed in Section 1.1. However, the metacharacters in this section create the greatest return on time investment due to how flexible and powerful they can make RegEx patterns.

Notice as we go through the examples how we can obtain some unexpected results. It is important to be very strategic when using some of these RegEx metacharacters as you don't always know what to expect in the text that you are processing. Even when you know the source quite well, there are inevitably errors or unknown changes that can wreck a poorly designed pattern. So, like any good analyst, you need to be thinking a few steps ahead in order to maintain robust RegEx code.

Note: Unlike SAS, all RegEx metacharacters are case sensitive, as you will see shortly. If a letter is defined here as lowercase or uppercase, then it MUST be used that way. Otherwise, your programs will do something very different from what you expect. In other words, even though you can be lazy with capitalization when writing SAS code (e.g., DATA vs. data), the same is not true here.

1.3.1 Wildcard

The wildcard metacharacter, which is a period (.), matches any single character value, except for a newline character (\n). The ability to match virtually any single character will prove useful when you are searching for the superset of associated character strings. You might also want to use it when you have no idea what values might be in a particular character position. Table 1.3 provides examples.

Table 1.3: Examples Using .

Usage	Matches
/R.n/	"Ran" "Run" "R+n" "R n" "R(n" "Ron" …
/.un/	"Fun" "fun" "Run" "run" "bun" "(un" "-un" …
/Street./	"Street." "Street," "Streets" "Street+" "Street_"…

Note: The period matches anything except the newline character (\n)—including itself. This can be helpful, but must be used wisely. Also note, only \n matches the newline character.

1.3.2 Word

The metacharacter \w matches any word character value, which includes alphanumeric and underscore (_) values. It matches any single letter (regardless of case), number, or underscore for a single character position. But do not be fooled by the underscore inclusion; \w does NOT match hyphens, dashes, spaces, or punctuation marks. Table 1.4 provides examples.

Table 1.4: Examples Using \w

Usage	Matches
/R\wn/	"Ran" "Run" "Ron" …
/\wun/	"Fun" "fun" "Run" "run" "Bun" "bun" "_un" …
/Street\w/	"Streets" "Street_"

Note: The \w wildcard should not have any unintentional spaces before or after it. Such spaces result in the pattern trying to match those additional spaces in addition to the \w. (This goes for any RegEx metacharacter.)

1.3.3 Non-word

The metacharacter \W matches a non-word character value (i.e., everything that \w doesn't include, except for the ever-elusive \n). The \W metacharacter is valuable when you are unsure what is in a character cell but you know that you don't want a word character (i.e., alphanumeric and _). Table 1.5 provides examples.

Table 1.5: Examples Using \W

Usage	Matches
/Washington\W/	"Washington." "Washington," "Washington;"…
/D\WC\W/	"D.C." "D,C." "D C." "D C " …
/Street\W/	"Street." "Street," "Street+" …

Note: You will continue to see lowercase and uppercase versions of these RegEx characters acting as near opposites, with some exceptions. It might not be overly clever, but does help simplify matters.

1.3.4 Tab

The metacharacter \t matches only the tab character in a string. Unlike the RegEx characters to follow, this metacharacter matches only the tab whitespace character. This is especially useful when the tab holds some special significance, such as when you are processing tab-delimited text files. Table 1.6 provides examples.

Table 1.6: Examples Using \t

Usage	Matches
/SAS\t/	"SAS "
/SAS\tInstitute\tInc/	"SAS Institute Inc"
/Street\t/	"Street "

Note: This metacharacter does not have an opposite (i.e., \T does not exist).

1.3.5 Whitespace

The metacharacter \s matches on a single whitespace character, which includes the space, tab, newline, carriage return, and form feed characters. You must include this when you are matching on anything in text that is separated by white space, and you are unsure of which will occur. Table 1.7 provides examples.

Table 1.7: Examples Using \s

Usage	Matches
/SAS\s/	"SAS " "SAS "
/SAS\sInstitute\sInc/	"SAS Institute Inc" "SAS Institute Inc"
/Street\s/	"Street " "Street "

Note: This form of the \s metacharacter matches only one whitespace character. We review how to find multiple matches in Section 1.5.2 because that is frequently needed when you are matching text.

1.3.6 Non-whitespace

The metacharacter \S matches on a single non-whitespace character—the exact opposite of \s. This metacharacter is often used to account for unexpected dashes, apostrophes, commas, and so on, that might otherwise prevent a match. Table 1.8 provides examples.

Table 1.8: Examples Using \S

Usage	Matches
/Leonato\Ss/	"Leonato's" "Leonatoas" "Leonato_s" …
/Washington\S/	"Washingtons" "Washington." "Washington," …
/Street\S/	"Street." "Street," "Streets" "Street+" "Street_"…

1.3.7 Digit

The metacharacter \d matches on a numerical digit character (i.e., 0–9). This RegEx metacharacter is probably the most straightforward one as it has a very narrow focus. Just remember that a single occurrence of \d is for only one character position in any text. In order to capture larger numbers (i.e., anything greater than 9), you have to build patterns with multiple occurrences of \d. Table 1.9 provides examples, but we discuss more sophisticated methods for accomplishing this later in the chapter. (See "Repetition Modifiers" in Section 1.5.2.)

Table 1.9: Examples Using \d

Usage	Matches
/\dst/	"1st" "9st" "4st" …
/10\d/	"101" "102" "103" …
/1-800-\d\d\d-\d\d\d\d/	"1-800-123-4567" "1-800-789-3456" …

Note: Just remember that even though your pattern might be correct, the data is not necessarily correct (4st and 9st don't make sense!).

1.3.8 Non-digit

The metacharacter \D matches on any single non-digit character. Again, this is the opposite of the lowercase metacharacter \d. This metacharacter matches on every value that is not a number. Table 1.10 provides examples.

Table 1.10: Examples Using \D

Usage	Matches
/1\D800\D123\D4567/	"1-800-123-4567" "1.800.123.4567" …
/1560\DWilson\DBlvd/	"1560 Wilson Blvd" "1560_Wilson_Blvd" …
/19\D\D\DStreet/	"19th Street" "19th.Street" "19…Street" …

1.3.9 Newline

The metacharacter \n matches a newline character. It is quite useful for some patterns to know that you have encountered a new line. For instance, you might be processing addresses in a text file, which often contain different pieces of information on different lines. Table 1.11 provides examples.

Table 1.11: Examples Using \n

Usage	Matches
/103 Pennsylvania Ave\. NW,\nWashington, DC 20216/	"103 Pennsylvania Ave. NW, Washington, DC 20216"
/<html tag>\n/	"<html tag> " …
/v\ne\nr\nt\ni\nc\na\nl\nt\ne\nx\nt/	"v e r t i c a l t e x t" …

Note: The test code does not enable us to actually try this metacharacter because it uses data lines, which is a feature of SAS that intentionally ignores newline characters when typed (i.e., pressing the Enter key just creates the start of a new data line in the SAS code window). For this reason, newline characters are not present in data lines for you to read and match on. But have faith, for now, that this one works as advertised. You will discover ways to process different text sources in the next chapter, enabling you to process newline characters.

1.3.10 Bell

The metacharacter \a matches an alarm "bell" character. The alarm character falls into a class of non-printing or invisible characters that are part of the ASCII character set. ASCII was developed long ago when operating systems used non-printing characters fairly extensively. Today, however, these characters are relatively uncommon, and most often occur only in files meant for computers to read rather than humans—since they are not displayed. When encountered, these characters generate an alarm tone, or "bell," on a computer's internal speaker. While they are often associated with errors, they can also be used to alert users that the end of a file or process has been achieved (e.g., in a system log file). You can use this metacharacter when you know to expect such a character in a source file. Table 1.12 provides examples.

Table 1.12: Examples Using \a

Usage	Matches
/\a END OF FILE/	"**BEL** END OF FILE"
/PROCESS COMPLETED SUCCESSFULLY\a/	"PROCESS COMPLETED SUCCESSFULLY **BEL**" ...
/\aERROR/	"**BEL**ERROR" ...

Note: Since the alarm character is a non-printing ASCII character, I am representing its location in the matching text with the BEL ASCII character. However, remember that such a code does not appear in our text.

1.3.11 Control Character

The metacharacter \cA-\cZ matches a control character for the letter that follows the \c. For example, \cF matches control-F in the source. This is one of several examples where you might be processing less-often-used file types (i.e., not a file meant for humans to read). Control characters, or non-printing characters, were once used extensively by transactional computing and telecommunications systems. These control characters, while not visible in most text editors, are still part of the ASCII character set, and can still be used by older systems in these regimes. For our examples in Table 1.13, we stick with the convention that is used for the alarm metacharacter above—the standard ASCII abbreviation is used despite the fact that they are never actually seen in text.

Table 1.13: Examples Using \cA-\cZ

Usage	Matches
/\cP/	**DEL** the non-printing **Data Link Escape** ASCII control character ^P
/\cB/	**STX** the non-printing **Start of Text** ASCII control character ^B
/\cBhello\cC/	**STX**hello**ETX** the non-printing **Start of Text** ASCII control character ^B followed by the character string "hello" and completed with the non-printing **End of Text** ASCII control character ^C

1.3.12 Octal

The metacharacter \ddd matches an octal character of the form *ddd*.[6] It is used to match on the octal code for an ASCII character for which you are searching. It can be especially useful when you need to find specific non-printing ASCII characters in a file. The default behavior by SAS is to return the ASCII character associated with this octal code in the results. Table 1.14 provides examples.

Table 1.14: Examples Using \ddd

Usage	Matches	Notes
/\s\041\s/	" ! "	This octal code translates to the ! ASCII character.
/\110\105\114\114\117/	"HELLO"	This series of octal codes translate to the "HELLO" string of ASCII characters.

Usage	Matches	Notes
/\s\007\011\s/	" BELTAB "	These octal codes translate to the two non-printing ASCII characters **BEL** and **TAB**. Refer to our discussion of the alarm metacharacter in Section 1.3.10 regarding characters that are not displayed.

Note: You will discover how to search for ranges of these values in the next section (Section 1.4). Also note that the largest ASCII value is decimal 127, octal 177, and hexadecimal 7F.

1.3.13 Hexadecimal

The metacharacter \xdd matches a hexadecimal character of the form *dd*.[7] The purpose of our implementation here is again not about searching through raw hexadecimal files, etc. We are using this to search for the hexadecimal code associated with the ASCII characters that we want in a source (manipulation of raw hex data sources is a different book). Table 1.15 provides examples.

Table 1.15: Examples Using \xdd

Usage	Matches	Notes
/\x2B/	"+"	This hexadecimal code translates to the + ASCII character.
/\x31\x2B\x31\x3D\x32/	"1+1=2"	These hexadecimal codes translate to the 1+1=2 ASCII characters.
/\x30\x30\x20\x46\x46/	"00 FF"	This is a reminder that we can match hexadecimal numbers stored in ASCII, and that they are not the same.

1.4 Character Classes

In addition to using the built-in RegEx characters to match patterns, users have the ability to create custom character matching. This capability is derived via different uses of [and] (square braces). The square braces essentially create a custom metacharacter, where the items contained between the opening brace and closing brace are possible match values for a single character cell. In addition to putting a list characters inside the braces, you can also include metacharacters. Each metacharacter discussed below includes an example, which includes the use of a metacharacter, and they all have the same match results. Just for fun, they are all identifying a hexadecimal number range present in the ASCII source file (stored as ASCII characters in the source file, but representing the range of possible hexadecimal values).

Note: Remember that some of the components discussed in this section are special characters that must be escaped with \ in order to be matched in isolation. Specifically, these characters are: ^, [, and].

1.4.1 List

The metacharacter [...] matches any one of the specific characters or metacharacters listed within the braces. Being able to define an unordered list of things that you want to appear in a space is very convenient, and can sometimes be more convenient than the metacharacters that identify broad classes of character types. Table 1.16 provides examples.

Table 1.16: Examples Using [...]

Usage	Matches
/[abcABC]/	"a" "b" "c" "A" "B" "C"
/[0173]/	"0" "1" "3" "7"
/[CcBbRr]at/	"cat" "Cat" "bat" "Bat" "rat" "Rat"
/[\dABCDEF]/	"0" "1" "2" "3" "4" "5" "6" "7" "8" "9" "A" "B" "C" "D" "E" "F"

1.4.2 Not List

The metacharacter [^...] matches one of anything not listed within the braces, except for the newline character. Sometimes it is easier to write down what we don't want rather than what we do. And for that reason, we might want to use this metacharacter. We can quickly identify the unwanted items and define them here. Table 1.17 provides examples.

Table 1.17: Examples Using [^...]

Usage	Matches
/[^abcABC]/	"d" "e" "f" …
/[^0173]/	"2" "4" "5" "6" "8" "9"
/[^Cc]at/	"fat" "Fat" "hat" "rat" "mat" "Hat" …
/[^\WGHIJKLMNOPQRSTUVWX YZabcdefghijklmnopqrstuvwxyz_]/	"0" "1" "2" "3" "4" "5" "6" "7" "8" "9" "A" "B" "C" "D" "E" "F"

1.4.3 Range

The metacharacter [...-...] matches anything that falls into a range of character values. In other words, case matters for letters listed in the braces. RegEx, and by extension SAS, understands the inherent order of letters and numbers. Therefore, we can define any range of numbers or letters to be matched by this metacharacter. Table 1.18 provides examples.

Table 1.18: Examples Using [...-...]

Usage	Matches
/[f-m]/	"f" "g" "h" "i" "j" "k" "l" "m"
/[1-9]/	"1" "2" "3" "4" "5" "6" "7" "8" "9"
/[a-cA-C]/	"a" "b" "c" "A" "B" "C"
/[\dA-F]/	"0" "1" "2" "3" "4" "5" "6" "7" "8" "9" "A" "B" "C" "D" "E" "F"

1.5 Modifiers

There are two significant things that you probably noticed as missing from the previous sections, which are worth further discussion here. First, all of the applicable metacharacters thus far have ignored letter case. In other words, \w, \S, \D, and . all match on a letter regardless of whether it is lowercase or uppercase. However, there are situations in which the case of a letter becomes important, but the letter itself is not known in advance.

Second, we can use a single match character as many times as we like, which creates additional fuzziness for our matches. However, there is a downside to just typing them out: *each occurrence must exist in order to match the pattern.* For instance, if the source text for the \D examples above contained "19thStreet" with no spaces, we'd never find it by using \D three times. And since the primary goal of the RegEx capability is to have automated text processing, we need a robust way to make this kind of matching more flexible.

Over the next two subsections (1.5.1 and 1.5.2), we will work through ways to overcome these limitations by using modifiers. There are two types of modifiers, case modifiers and repetition modifiers. Combining them gives us significant robustness and flexibility in real-world RegEx implementations, and should be considered as fundamental to real-world implementations as the metacharacters that we have discussed thus far.

1.5.1 Case Modifiers

When performing matches on text, there is the obvious consideration of letter case (upper vs. lower). Although I have already introduced a rudimentary way to handle this in situations where the letter is known, there still must be a methodology for accounting for letter case when it is unknown. This section discusses a variety of approaches to dealing with case matching. Depending on the situation, some approaches are more convenient than others, while not necessarily being right or wrong.

Lowercase

The metacharacter \l matches when the next character in a pattern is lowercase. This metacharacter applies only to characters (metacharacters, groups, and so on, don't work). In practice, it is more practical to simply type the lowercase version of the desired character value, or provide a list of lowercase letters to match. Table 1.19 provides examples.

Table 1.19: Examples Using \l

Usage	Matches
/\lStreet/	"street" …
/\s\lS\lA\lS\sInstitute/	" sas Institute" …
/(\lS|\lF)leet/	"sleet" "fleet" …

Uppercase

The metacharacter \u matches when the next letter in a pattern is uppercase. It functions exactly as the lowercase version introduced above (\l), but also applies to uppercase. Table 1.20 provides examples.

Table 1.20: Examples Using \u

Usage	Matches
/\uinc./	"Inc." …
/\ustreet|\ust\./	"Street" "St." …
/\uave\.|\uavenue,/	"Ave." "Avenue,"

Lowercase Range

The metacharacter \L…\E matches when all the characters between the \L and \E are lowercase. Strings typed between \L and \E are forced to match on lowercase only, even when they are typed in as capital letters. However, unlike the \l metacharacter, \L…\E can also contain character classes and repetition modifiers. Table 1.21 provides examples.

Table 1.21: Examples Using \L…\E

Usage	Matches
/\L[a-z0-9][a-z0-9][a-z0-9]\E/	"sas" "abc" "123" …
/\LTHESE ARE LOWERCASE\E/	"these are lowercase"
/\sR\L[a-z][a-z][a-z]\E\s/	" Read " " Road " " Rode " " Ride " " Real " …

Note: When applying case modifiers to non-alphabet characters, the modifier is ignored. It doesn't apply to those characters, so it doesn't affect the match.

Uppercase Range

The metacharacter \U…\E creates a match when all the characters between the \U and \E are uppercase. Again, this metacharacter functions the same way as the lowercase version discussed above, but applies to uppercase. This metacharacter can be useful for identifying acronyms or other text where capital letters are important. Table 1.22 provides examples.

Table 1.22: Examples Using \U…\E

Usage	Matches
/\U[a-z][a-z][a-z]\E/	"SAS" "CIA" …
/\U[a-z][a-z][a-z]\E\sInstitute\sInc\W/	"SAS Institute Inc." …
/\s\Uallcaps\E\s/	" ALLCAPS "

Note: Notice that other metacharacters are not allowed inside \L…\E or \U…\E metacharacters. In other words, \w can't be used to replace the character classes above.

Quote Range

The metacharacter \Q...\E matches all content inside the \Q and \E as character strings, disabling everything including the backslash character. Metacharacters cannot be used inside \Q...\E. The functionality provided by this metacharacter is great for searching within strings that contain a significant number of reserved characters, such as XML, webserver logs, or HTML. Table 1.23 provides examples.

Table 1.23: Examples Using \Q...\E

Usage	Matches
/\Q<html tag name>\E/	"<html tag name>"
/\Qf(x) + f(y) = z\E/	"f(x) + f(y) = z"
/\Q<!DOCTYPE HTML> <html lang="en-US">\E/	"<!DOCTYPE HTML> <html lang="en-US">"

1.5.2 Repetition Modifiers

Repetition modifiers change the matching repetition behavior of the metacharacters and characters immediately preceding them in a pattern. They can also modify the matching repetition of an entire group—defined using () to surround the group of metacharacters and characters before the modifier. Just keep in mind that repetition of the entire group means that it repeats back-to-back (e.g., "haha"), unless we also modify the individual metacharacters.

Now, there are two types of repetition modifiers, *greedy* and *lazy*. Greedy repetition modifiers try to match as many times as possible within the confines of their definition. Lazy modifiers attempt to find a match as few times as possible. They have similar uses, which can make the difference between their results subtle.

Introduction to Greedy Repetition Modifiers

Let's start by discussing greedy modifiers because they are a little more intuitive to use. As we go through the examples, it is important to keep in mind that greedy modifiers match as many times as possible—constantly searching for the last possible time the match is still true. It is therefore easy to create patterns that match differently from what you might expect.

There is a concept in RegEx known as *backtracking*, which is the root cause for potential issues with greedy modifiers (hint: backtracking results in the need for lazy modifiers). As we discuss further when we examine lazy repetition modifiers, a greedy modifier actually tries to maximize the matches of a modified pattern chunk by searching until the match fails. Upon that failure, the system then *backtracks* to the position where the modified chunk last matched. The processing time wasted with backtracking for a single match is insignificant. However, as soon as we introduce a few additional factors, this problem can waste tremendous computing cycles—multiple modified pattern chunks, numerous match iterations (think loops), and large data sources. It is important to be mindful of these factors when designing patterns as they can have unintended consequences.

Greedy 0 or More

The modifier * requires the immediately preceding character or metacharacter to match 0 or more times. It enables us to generate unlimited optional matches within text. For example, we might want to match every occurrence of a word root, along with all of its prefixes and suffixes. By allowing the prefixes and suffixes to be optional, we are able to achieve this goal. Table 1.24 provides examples.

Table 1.24: Examples Using *

Usage	Matches
/Sing\w*/	"Sing" "Sings" "Singing" "Singer" "Singers" …
/D\W*C\W*/	"DC" "D.C." "D C " "D….-!$%^ C.-)*&^%"…
/19\D*Street/	"19th Street" "19thStreet" "19Street" …
/Hello*/	"Hell" "Hello" "Helloooooooooooooo" …

Greedy 1 or More

The modifier + requires the immediately preceding character or metacharacter to match 1 or more times. The plus sign modifier works similarly to the asterisk modifier, with the exception that it enforces a match of the metacharacter or character at least 1 time. Table 1.25 provides examples.

Table 1.25: Examples Using +

Usage	Matches
/Ru\w+/	"Run" "Ruin" "Runt" "Runners" …
/\s\U[a-z]+\E\s/	Words with all letters capitalized, and surrounded by spaces.
/19\D+Street/	"19th Street" "19th.Street" "19…Street" …
/(ha)+/	"ha" "hahahahahahaha" …

Note: Pay special attention to the addition of the \s metacharacter in the second example in Table 1.25. If it were not present, the pattern would also match only single capital letters at the beginning of words. By adding \s, the pattern requires a whitespace character to immediately follow the one or more capital letters, thus eliminating matches on single letters at the beginning of words.

Greedy 0 or 1 Time

The modifier ? creates a match of only 0 or 1 time. The question mark provides us the ability to make the occurrence of a metacharacter optional without allowing it to match multiple times. This can be effective for matching word pairs that have inconsistent use of dashes or spaces (e.g., short-term vs. short term). Table 1.26 provides examples.

Table 1.26: Examples Using ?

Usage	Matches
/1\D?800\D?123\D?4567/	"1-800-123-4567" "18001234567" …
/1560\sWilson\sBlvd\W?/	"1560 Wilson Blvd." "1560 Wilson Blvd" …
/19th\s?Street/	"19th Street" "19thStreet" …

Greedy n Times

The modifier {n} creates a match of exactly *n* times. Being able to match on a metacharacter exactly *n* number of times is the same as typing that metacharacter out that many times. However, from the perspective of coding and maintaining the RegEx patterns, using the modifier is a much better approach. It limits the opportunity for us to make typographical errors when initially creating the RegEx pattern, and it improves readability when later editing and sharing the patterns. Table 1.27 provides examples.

Table 1.27: Examples Using {n}

Usage	Matches
/1-800-\d{3}-\d{4}/	"1-800-123-4567" "1.800.123.4567" ...
/R\w{4}/	"Round" "Runts" "Ruins" ...
/19\D{3}Street/	"19th Street" "19th.Street" "19...Street" ...
/(\d{5}-\d{4})|(\d{5})/	"12345-6789" "12345" ...

Greedy n or More

The modifier {n,} creates a match at least *n* times. By ensuring that we can match something at least *n* times, we are able to create functionality very similar to the plus modifier. However, we are raising the minimum number of times that the metacharacter must match. This is quite useful for certain applications, but must be handled with caution. Also, like the + modifier, we can easily get very long strings of unanticipated matches due to a single logical error in pattern construction. Table 1.28 provides examples.

Table 1.28: Examples Using {n,}

Usage	Matches
/1-800-\d{1,}-\d{2,}/	"1-800-123-4567" "1-800-789-12" ...
/\d{3,}-\d{2,}-\d{4,}/	"143-25-7689" "12345689-546545654-9820"...
/19\D{3,}Street/	"19th Street" "19th, Not My Street" ...

Note: Be mindful not to type a space after the comma inside the curly braces. It is easy to do out of habit, but it will wreck our pattern!

Greedy n to m Times

The modifier {n,m} creates a match at least *n*, but not more than *m* times. Creating a match with a specified range is quite useful for ensuring that data quality standards are being maintained. When extracting semi-structured data elements such as ZIP codes, birthdates, and phone numbers, it is important to maintain a certain level of flexibility while also ensuring that the source is within expected tolerances. For instance, a two-digit year might be accepted in lieu of a four-digit year, but a four-digit zip would be unacceptable. Table 1.29 provides examples.

Table 1.29: Examples Using {n,m}

Usage	Matches
/(1-)?8\d\d-\d{3,3}-\d{4,4}/	"1-800-123-4567" …
/\d{1,2}-\d{1,2}-\d{2,4}/	"10-20-1950" "8-30-52" "4-3-1979"…
/Was{1,7}/	"Washington" "Wash" "Waste" "Washing" …

Note: As you can see in the examples above, the {n,m} might not always be the best choice of modifier, but these examples are meant to demonstrate the flexibility of implementation. For instance, the year in the second example is allowed to be three digits with this usage. Using an OR clause with the {n} modifier is a simple fix.

Introduction to Lazy Repetition Modifiers

Now that you are familiar with greedy modifiers, let's begin examining the lazy ones. In terms of syntax, they differ from the greedy modifiers only by the addition of a question mark (?). By adding the question mark immediately after each of the greedy modifiers, we are able to subtly change their behavior—sometimes in unexpected ways.

In general, lazy modifiers are used to both avoid overmatching and improve performance when compared to the greedy modifiers. There are situations when matching with greedy modifiers would lead to either grabbing too much information, or simply slowing down system performance. For instance, processing semi-structured text files such as HTML or XML is a great example of when lazy modifiers would come in handy.

Lazy 0 or More

The modifier *? creates a match 0 or more times, but as few times as necessary to create the match. In some situations, it creates the same matches as does the greedy version. However, in other cases, the results are very different. To make it clearer, Table 1.30 describes the details of a few examples.

Table 1.30: Examples Using *?

Usage	Matches	Notes
/Sing\w*?/	"Sing"	This matches only the word "Sing" because the modifier is given the option to match nothing. And since it is *lazy*, it will take that option every time, regardless of whether a word character immediately follows the "g" in "Sing".
/Sing\w*?\s/	"Singing " …	Comparing this to the example above, you see that appending the \s on the pattern creates additional matches. The \s forces the pattern to continue searching for a match that includes white space. This could be "Sing " or many other combinations (similar to the greedy outcomes).
/(ha)*?/	""	This example demonstrates why we need to be careful with lazy modifiers. Even when "ha" exists, it is ignored, again because the modifier has the option to do so. The greedy version of this would match as many times as the word "ha" occurred back-to-back, with a minimum of zero times.

Lazy 1 or More

The modifier +? creates a match 1 or more times, but as few times as necessary to create a match. Again, if it is possible, this matches only once. Table 1.31 provides examples.

Table 1.31: Examples Using +?

Usage	Matches	Notes
/Sing\w+?/	"Singi"	This matches only "Sing" plus exactly one word character following the "g". Again, by giving the lazy modifier an option to match the minimum, it will do so every time.
/Sing\w+?\s/	"Singing " …	Again, we see that appending the \s on the pattern creates additional matches. The \s forces the pattern to continue searching for a match that includes white space. This could be "Singi " or many other combinations (similar to the greedy outcomes).
/(ha)+?/	"ha"	This example is less of a cautionary tale than for *?. But it might still provide undesirable results. Even when "ha" exists numerous times back-to-back, it matches only the first time, unless an additional match element follows it. Again, this is because the modifier has the option to match only once. The greedy version of this would match as many times as the word "ha" occurred back-to-back, with a minimum of once.

Lazy 0 or 1 Times

The modifier ?? creates a match 0 or 1 times, but as few times as necessary to create a match. Unless forced, this modifier will match 0 times. Table 1.32 provides examples.

Table 1.32: Examples Using ??

Usage	Matches	Notes
/Sing\w??/	"Sing"	This matches only the word "Sing" because the modifier is given the option to match nothing. And since it is *lazy*, it will take that option every time, regardless of whether a word character immediately follows the "g" in "Sing". The reasoning is the same as with the *? modifier.
/Sing\w??\s/	"Sings " …	Again, just as with the *? modifier, we see that appending the \s on the pattern creates additional matches. The \s forces the pattern to continue searching for a match that includes white space. This could be "Sings " or a few other combinations (similar to the greedy outcomes).
/(ha)??/	""	This example demonstrates why we need to be careful with lazy modifiers. Even when "ha" exists, it is ignored, again because the modifier has the option to do so. The greedy version of this would match as many times as the word "ha" occurred back-to-back.

Lazy n Times

The modifier {n}? creates a match exactly *n* times. This modifier functions exactly as the greedy version, making the ? unnecessary. Using this modifier results in no performance enhancement or change in functionality, which makes it a completely unnecessary addition to the Perl language. It has been included here for the sake of completeness. Table 1.33 shows that the same examples reveal the same results.

Table 1.33: Examples Using "{n}?"

Usage	Matches
/1-800-\d{3}?-\d{4}?/	"1-800-123-4567" "1.800.123.4567" …
/R\w{4}?/	"Round" "Runts" "Ruins" …
/19\D{3}?Street/	"19th Street" "19th.Street" "19…Street" …
/(\d{5}?-\d{4}?)|(\d{5}?)/	"12345-6789" "12345" …

Lazy n or More

The modifier {n,}? creates a match, at least *n* times and as few times as necessary to create a match. This functions just like the *? or +? modifiers, except that the minimum number of matches is arbitrary. Again, we see similar behavior resulting from the laziness of the modifier. Table 1.34 provides examples.

Table 1.34: Examples Using {n,}?

Usage	Matches	Notes
/Sing\w{3,}?/	"Singing" …	This usage matches exactly *n*=3 times. Again, by giving the lazy modifier an option to match the minimum, it will do so every time.
/0{3,}?\s/	"0000 " …	Now that you have the hang of these modifiers, this example should be a little more interesting. Appending \s on the pattern still forces it to match each 0 until the white space is encountered. The pattern is "anchored" to the first occurrence of a 0, thus capturing more than the minimum.
/(ha){4,}?/	"hahahaha"	Without surrounding information in the pattern, this matches only the minimum number of times. By having nothing else to force additional matching, the lazy modifier just stops after the minimum of *n*=4.

Lazy n to m Times

The modifier {n,m}? creates a match at least *n* times, but no more than *m* times—as few times in that range as necessary to create the match. It functions like many of the other lazy modifiers discussed thus far, but it sets a cap on how many times it can match in addition to having an arbitrary minimum. Table 1.35 provides examples.

Table 1.35: Examples Using {n,m}?

Usage	Matches	Notes
/Read\w{1,3}?/	"Ready" …	This usage matches the word metacharacter only one time. Again, by giving the lazy modifier an option to match the minimum, it will do so every time.
/0{2,5}?\s/	"0000 " …	Again, the pattern is "anchored" to the first occurrence of a 0, thus capturing the minimum if it exists, up to the maximum.
/\sha(ha){0,6}?/	" ha"	By not having anything after the "anchor" point for the pattern to match on, there is nothing to force additional matching. The lazy modifier just stops after the minimum of *n*=0.

1.6 Options

Options affect the behavior of the entire RegEx pattern with which they are associated. These behavioral changes provide benefits ranging from making RegEx creation more convenient, to providing new or enhanced functionality.

Options occur *after* the closing slash character, but there is one item of significance that occurs *before* the first slash character that we will also discuss—it is not actually an option but this is best place to go over it. And we are not going to cover all of the options for the same reason we haven't covered absolutely all of the metacharacters thus far—this is an introductory text.

1.6.1 Ignore Case

The option //i ignores letter case for the entire pattern, even character strings. This is a great option to use when we know exactly what words we are searching for, but we don't want the letter case to be an issue. Table 1.36 provides examples.

Table 1.36: Examples Using //i

Usage	Matches
/1600 Pennsylvania Avenue/i	"1600 pennsylvania avenue" "1600 PENNSYLVANIA AVENUE" …
/STREET/i	"street" "Street" "STREET" …
/CAPS don't MaTtEr/i	"caps don't matter" "CaPs DoN't MATTER" …

1.6.2 Single Line

The option //s forces the dot character (.) to match everything, including the newline character, when it occurs in the pattern. This can be very helpful to ensure that we don't miss anything for a particular character position. Table 1.37 provides examples.

Table 1.37: Examples Using //s

Usage	Matches
/43rd and Times Square.New York, NY 10036/s	"43rd and Times Square New York, NY 10036" …
/Bob Smith.\d{3}-\d{3}-\d{4}/s	"Bob Smith 123-456-7891" …

1.6.3 Multiline

The option //m causes ^ and $ to match on more than just the string start and end respectively. Instead, they match on every newline encountered because the various lines of information are treated as one continuous line. This enhanced functionality really applies to two metacharacters that we haven't covered yet (we'll discuss them in Section 1.7), so if you need to, feel free to peek ahead and come back to this one. Table 1.38 provides examples.

Table 1.38: Examples Using //m

Usage	Matches
/^\w+/m	Words at the beginning of a string and words following a newline character.
/\w+?\s$/m	Words immediately before a space and the string end, and before a space and newline character.

1.6.4 Compile Once

The option //o is known as the *compile once* option. By having the "o" immediately following the closing slash, SAS knows to compile that RegEx only once. This option creates a very nice simplification to SAS code, which I demonstrate by showing updated test code below (see Section 1.1.4 for the original code). Notice how the IF block is removed, and only the two lines that do not include the RETAIN statement remain. These changes are possible due to the compilation happening the first time through the DATA step. Every subsequent loop through reuses the previously compiled expression, if it exists.

Updated Test Code

```
/*RegEx Testing Framework*/
data _NULL_;
*if _N_=1 then
*do;
*    retain pattern_ID;
*    pattern="/Run/"; /*<--Edit the pattern here.*/
*    pattern_ID=prxparse(pattern);
*end;
```

```
pattern="/Run/o"; /*<--Edit the pattern here.*/
pattern_ID=prxparse(pattern);
input some_data $50.;
call prxsubstr(pattern_ID, some_data, position, length);
if position ^= 0 then
   do;
      match=substr(some_data, position, length);
      put match:$QUOTE. "found in " some_data:$QUOTE.;
   end;
datalines;
Smith, BOB A.
ROBERT Allen Smith
Smithe, Cindy
103 Pennsylvania Ave. NW, Washington, DC 20216
508 First St. NW, Washington, DC 20001
650 1st St NE, Washington, DC 20002
3000 K Street NW, Washington, DC 20007
1560 Wilson Blvd, Arlington, VA 22209
1-800-123-4567
1(800) 789-1234
;
run;
```

1.6.5 Substitution Operator

While the substitution operator s// is not technically an option, it belongs here if only because it truly stands apart from the other items discussed in this section. Although the substitution operation is similar in appearance to the other options, it fundamentally changes the RegEx activity from a matching operation to a match-and-replace operation. Placing "s" in front of the surrounding slashes (//) signifies that the pattern is being used to replace the text being matched and insert the accompanying replacement text. This operator is another peek at additional functionality that is explored in the next chapter with SAS functions. Once a pattern is matched, we can then do a variety of things with that information. A great analogy for how this works in practice is the find-and-replace functionality provided by many word processing applications—except this is much more powerful. Also, notice that there is a third slash in the examples below (in the middle of the patterns). That additional slash denotes where the matching portion of the RegEx ends and the replacement portion begins. And notice something important in the last example: *everything is a string literal*. That's right, all the characters that occur between the second and third slash are treated as just characters. Table 1.39 provides some examples, but we will cover this in detail in the next chapter, where we also discuss how to insert more than just character strings.

Table 1.39: Examples Using s//

Usage	Matches	Replaces with
s/Stop/Go/	"Stop"	"Go"
s/Sing/Read/	"Sing"	"Read"
s/1\s?\(800\)\s?-\s?/1-800-/	"1 (800) - " …	"1-800-"

Note: This is a more advanced function that our test code is not set up to handle. You'll just need to accept it as true until we use it with some SAS code in the next chapter.

1.7 Zero-width Metacharacters

Zero-width characters, often called positional characters, are not matched in isolation because they do not have a width. They are used as an additional piece of information for making a proper pattern match. There are numerous examples for how these zero-width characters can be used. For instance, perhaps you want to match a particular word, but only if it occurs at the beginning of a line.

1.7.1 Start of Line

The metacharacter ^ matches the beginning of a line or string. Depending on the text that we are processing, we might know a priori that a new line signifies something specific. For example, we might be looking for the beginning of a new paragraph, which could be denoted by a new line in combination with a capital letter and no preceding white space. Or we might need to be prepared to match an address that includes a new line for the city, state, and zip. Table 1.40 provides examples.

Table 1.40: Example Using ^

Usage	Matches
/^Washington, DC 20007/	" Washington, DC 20007"
/^\w+\b/	The first word in a string.

Note: This metacharacter is often used as the logical NOT symbol, including within the character class metacharacters discussed in Section 1.3 and in SAS code. So be careful not to get confused in its usage when shifting between contexts.

1.7.2 End of Line

The metacharacter $ matches the end of a line or string. There are numerous situations in which this might become relevant, similar to the reasons for the ^ metacharacter. Table 1.41 provides examples.

Table 1.41: Example Using $

Usage	Matches
/3000 K Street NW,$/	"3000 K Street NW, "
/\$\d+?\.\d{2}\s*?$/	"$150.52 "

1.7.3 Word Boundary

The metacharacter \b matches a word boundary. The \b RegEx assertion metacharacter is zero-width because it actually represents the invisible gap between two characters, with a \w character on one side and \W on the other. Therefore, when you use this metacharacter, you won't generate matches that contain the associated non-word character. Table 1.42 provides examples.

Table 1.42: Example Using \b

Usage	Matches
/Street\b/	"Street" from the substrings, "Street," "Street " …
	But does NOT match from the substring "Streets" etc.
/\b8\d{2}\b/	"800" "888" … from the substrings "(800)" "-888-" …
	But does NOT match from the substrings "18002" …
/\b\U[a-z]+\E\b/	Words in all caps. Without the second \b, the output would also include single capitalized letters from the front of a word.

1.7.4 Non-word Boundary

The metacharacter \B matches a non-word boundary (i.e., anywhere \b does not match). This is especially useful for matching root words or substrings without including the surrounding pieces of information. Table 1.43 provides examples.

Table 1.43: Examples Using \B

Usage	Matches
/read\B/	"read" from the substrings, "reads" "reading" "reader" …
	But does NOT match from the substring "read"
/\Bun\b/	"un" from the substrings, "fun " "rerun." "gun," …
	But does NOT match from the substring "un"
/\b[a-zA-Z]{3,}\b/	Any word longer than three letters.

1.7.5 String Start

The metacharacter \A matches the beginning of a string. Similar to the word boundary metacharacter (\b), \A occurs between two character cells. It also denotes when a string value occurs to its right with nothing to its left. In the context of data lines (as in our test code for this chapter), that situation occurs at the beginning of each line.

However, suppose we had a more complex task such as stitching together multiple strings of extracted text (stored in SAS variables). In this context, \A could be a key to determining in what order to place or sort them. However, for our test code, the \A matches only on the beginning of each data line, since each line is identified as the beginning of the string. So, this is another one that you have to approach with a little bit of faith until we start doing some more interesting tasks in the next chapter. Table 1.44 provides examples.

Table 1.44: Examples Using \A

Usage	Matches
/\A\w*?\s/	The first word of a line. In the case of our test code, it matches: "ROBERT " from line 2; "103 " from line 4; "508 " from line 5; "650 " from line 6; "3000 " from line 7; and "1560 " from line 8.

1.8 Summary

We have explored a variety of interesting new concepts in this chapter, and I've been doing my utmost to make them tangible along the way. Hopefully, you are now ready to tackle the challenge of implementing these concepts in SAS code in the coming chapters. Following are some takeaways that you should keep in mind for the coming pages and beyond.

Flexibility

It should have become clear through reading this chapter that there are many ways to accomplish the same task, making few of them truly right or wrong. You have to decide the most efficient and effective approach for accomplishing your goals to determine what is best for a given situation.

Scratching the Surface

We have only begun to scratch the surface of what RegEx can do. The information that you have learned thus far is a solid foundation upon which you can develop sophisticated functionality.

Start Small

As we have explored a variety of RegEx capabilities throughout this chapter, it is easy to become overwhelmed with attempting to do too much at once. As with anything, it is best to start small by experimenting with simple patterns and iteratively evolve them. And remember that leveraging just a few of the elements that we have covered can have a tremendous impact on the processing and analysis of textual information.

1 Wikipedia contributors, "Regular expression," *Wikipedia, The Free Encyclopedia*, https://en.wikipedia.org/w/index.php?title=Regular_expression&oldid=857059914 (accessed August 29, 2018).

2 For more information on the version of Perl being used, refer to the artistic license statement on the SAS support site here: http://support.sas.com/rnd/base/datastep/perl_regexp/regexp.compliance.html

3 SAS Institute Inc. "Tables of Perl Regular Expression (PRX) Metacharacters," SAS 9.4 Functions and CALL Routines: Reference, Fifth Edition, http://support.sas.com/documentation/cdl/en/. lefunctionsref/67239/HTML/default/viewer.htm#p0s9ilagexmjl8n1u7e1t1jfnzlk.htm (accessed August 29, 2018).

4 "perlre," *Perl Programming Documentation,* http://perldoc.perl.org/perlre.html (accessed August 29, 2018).

5 SEC, "SEC Administrative Proceedings for 2009," *U.S. Securities and Exchange Commission,* http://www.sec.gov/open/datasets/administrative_proceedings_2009.xml (accessed August 29, 2018).

6 Octal is a number system that uses base-8 instead of base-10. This system has only numbers 0–7 represented. Some old microcontrollers and microprocessors used this encoding, but it is extremely rare today.

7 Hexadecimal is a number system that uses base-16 instead of base-10. The possible values go from "0" to "F" in a single character position (where A=10, B=11, ..., F=15).

Chapter 2: Using Regular Expressions in SAS

2.1 Introduction

This chapter is focused on developing your understanding of built-in SAS functions and call routines, and on starting to do some real SAS coding. Here, you will learn the mechanics of how to implement the wonderful RegEx metacharacters introduced in Chapter 1. Each function or call routine introduced has associated examples to ensure that their use is clear. We also briefly discuss how each is useful.

2.1.1 Capture Buffer

Now, before we go any farther, we have to address a concept called the *capture buffer*. The capture buffer is a more advanced technique that I have avoided delving into thus far, but it must be understood so that you can use some functions (required for PRXPAREN and PRXPOSN, but optional for PRXCHANGE). As you recall from Chapter 1, parentheses create logical groupings within a RegEx, but they also do something more interesting. For every set of parentheses used in a particular RegEx pattern, a slot in a memory buffer is created. This slot in memory is then referenceable just like any variable (a more experienced programmer can think of it like a pointer buffer). Each slot is created in sequential order of parentheses pair occurrence and is referenced accordingly using the $ sign.

For example, the RegEx s/(The) (cat) (is) (fat)/$4 $3 $1 $2/ creates the output "fat is The cat". Now, imagine applying that same ability to unknown data elements instead of just to string literals. This could become a very powerful capability for standardizing or restructuring data to meet specific needs.

2.2 Built-in SAS Functions

In this section, we cover the SAS functions for performing RegEx operations. SAS functions for RegEx have the same usage limitations as other built-in functions. (See SAS documentation.) Also just like all other functions, they can only take arguments and return output in assignment statements and expressions.

Note: Each RegEx function has PRX at the beginning, which represents Perl-Regular-eXpressions.

2.2.1 PRXPARSE

Description

This function takes a RegEx pattern as input and provides a numerical RegEx pattern identifier as output. The unique pattern identifier is used by other functions and call routines to reference the pattern. This function should look familiar since we used it in our example code in Chapter 1.

Syntax
```
RegEx_ID = PRXPARSE (RegEx)
```

RegEx: The pattern to be parsed (input argument, required)

RegEx_ID: Unique numerical RegEx identifier returned by PRXPARSE (output, required)

Now, it is important to understand at this point that PRXPARSE compiles the RegEx in order to create the identifier for SAS to later reference and use. And this is what makes the RegEx //o option so important when using PRXPARSE in code. The //o option forces SAS to compile the RegEx code once, creating the RegEx identifier the first time only. When a particular RegEx is intended to be reused on every loop through the DATA step, we want to leverage this functionality in order to avoid recompiling the RegEx pattern every time it is encountered in code (i.e., on each iteration of a DATA step). If the pattern is not definitely going to be used every time through the DATA step (e.g., it's not defined inside an IF statement), then we might not want to waste memory maintaining it. In other words, we might not always want to use the //o option—the decision is about tradeoffs. When you're dealing with very few of these patterns or with a small amount of data, the tradeoffs don't really apply. But when we scale up to a system using hundreds of patterns, or tens of millions of records, the tradeoffs (speed at the expense of memory usage) become very important.

Example 2.1: Defining Patterns with PRXPARSE

Let's revisit the last bit of example code from Chapter 1 since it is already familiar. The RegEx below is defined as /Smith/o, meaning that we are looking for any occurrence of the string literal "Smith" within the data lines provided. This RegEx is the argument for PRXPARSE, which creates a pattern identifier that is assigned to the variable pattern_ID. This variable is then passed to the call routine PRXSUBSTR, which is discussed in section 2.3. Because you are familiar with the overall function of this code by now, this need not be a distraction.

The output of this RegEx is presented in Output 2.1. As we expected, the code found every occurrence of "Smith" regardless of what was surrounding it—including other letters.

```
/*RegEx Testing Framework*/
data _NULL_;
pattern = "/Smith/o"; /*<--Edit the pattern here.*/
```

```
pattern_ID = PRXPARSE(pattern);
input some_data $50.;
call prxsubstr(pattern_ID, some_data, position, length);
if position ^= 0 then
    do;
        match = substr(some_data, position, length);
        put match:$QUOTE. "found in " some_data:$QUOTE.;
    end;
datalines;
Smith, BOB A.
ROBERT Allen Smith
Smithe, Cindy
103 Pennsylvania Ave. NW, Washington, DC 20216
508 First St. NW, Washington, DC 20001
650 1st St NE, Washington, DC 20002
3000 K Street NW, Washington, DC 20007
1560 Wilson Blvd, Arlington, VA 22209
1-800-123-4567
1(800) 789-1234
;
run;
```

Output 2.1: Log Output of Pattern /Smith/o

```
"Smith" found in "Smith, BOB A."
"Smith" found in "ROBERT Allen Smith"
"Smith" found in "Smithe, Cindy"
```

2.2.2 PRXMATCH

Description

PRXMATCH returns the numerical position of the first character in the matched RegEx pattern. In addition, it can be used in IF statements to test for a pattern match without a variable assignment, just like many other familiar SAS functions. The first argument to PRXMATCH is either the RegEx or RegEx_ID. The second is the source text variable or string literal.

Syntax

```
Position = PRXMATCH(RegEx_ID or RegEx, Source_Text)
```

RegEx_ID: Unique RegEx identifier returned by PRXPARSE (input argument, required if RegEx not used)

RegEx: The pattern to be matched (input argument, required if RegEx_ID not used)

Source_Text: The text variable or literal to be operated upon (input argument, required)

Position: Numerical position variable assignment (output, required)

As we discussed with the PRXPARSE function, RegEx patterns are compiled by SAS for use by other functions. Therefore, in addition to using the actual RegEx, PRXMATCH is able to leverage the previously compiled RegEx via the RegEx_ID argument in lieu of the RegEx itself. This allows us to compile the

RegEx once via the PRXPARSE function (using the //o option), minimizing the associated computing cycles. Such small savings in computing cycles can prove significant when processing large volumes of text.

The two different methods for leveraging RegEx patterns create significant flexibility in how PRXMATCH can be used in practice. By not needing to compile the RegEx in advance, PRXMATCH allows us to embed RegEx patterns throughout our code without the extra memory allocation required to maintain them for each loop through the DATA step. This is very useful when you are using PRXMATCH in a dynamic way, such as inside nested IF statements where the RegEx is used only when certain conditions are true. Depending on the implementation, there are implications for speed as well as for memory usage.

Example 2.2: Finding Strings in Source Text with PRXMATCH

Let's try a simple example to see how this function is used in practice. Suppose we want to find a string such as "Street" in a source text. The code below demonstrates how we print the position of each occurrence to the log. Obviously, we need to do more than just print the position in practice (such as by extracting or manipulating the matched text), but this demonstrates the basic functionality of PRXMATCH.

The PRXMATCH function is implemented in this code with the RegEx as the first argument and the `datalines` reference `address` as the second argument. The result is assigned to the variable Position. The value of Position is then written to the log using the PUT statement.

Count the character positions in the data lines. At what position do we encounter the "S" in "Street" on the lines in which they occur? Comparing the results to Output 2.2, we see that PRXMATCH is returning the position of "S" (i.e., the position for the first character in the pattern match).

```
data _NULL_;
input address $50.;
position = PRXMATCH('/Street/o', address);
if position ^= 0 then
    do;
        put position=;
    end;
datalines;
103 Pennsylvania Ave NW, Washington, DC 20216
508 First Street NW, Washington, DC 20001
650 1st St NE, Washington, DC 20002
3000 K Street NW, Washington, DC 20007
;
run;
```

Output 2.2: Log Printout for Positions of "Street"

```
position=11
position=8
```

2.2.3 PRXCHANGE

Description

This function searches for the pattern—provided in the first argument by either RegEx_ID or RegEx—within the source text that is provided in the third argument. The pattern is matched the number of times given in the second argument, Num_Times. Upon finding each match, the function then returns the changed text as required by the RegEx. If no match is found, PRXCHANGE returns the original text unchanged.

Syntax

```
Output_String = PRXCHANGE(RegEx_ID or RegEx, Num_Times, Input_String)
```

RegEx_ID: Unique RegEx identifier returned by PRXPARSE (input argument, required if RegEx not used)

RegEx: The pattern to be matched (input argument, required if RegEx_ID not used)

Num_Times: Number of times the change is to be applied (input argument, required). -1 forces the function to make the changes as many times as the pattern occurs in the source text.

Input_String: Input text variable (input argument, required)

Output_String: Output text variable assignment (output, required)

Just like the PRXMATCH function, PRXCHANGE is able to use the actual RegEx pattern or the RegEx_ID, providing significant flexibility. The preferred use again depends on the desired application.

This function is very useful for data standardization, as you will see in more advanced examples in the next chapters. We will work through two examples below to demonstrate some more basic functionality of the PRXCHANGE function, as well as to demonstrate how to leverage the capture buffer concept introduced earlier.

Example 2.3: Standardizing Data

Data standardization is a relatively simple, yet powerful, capability provided by RegEx in SAS. PRXCHANGE enables us to scrub our data source to ensure that each occurrence of a word or phrase is exactly the same (or removed entirely). See Output 2.3 for the results. Data scrubbing becomes especially important when you are attempting to perform advanced applications such as text mining.

For instance, before doing any analysis of our data, we want to know that each occurrence of the word "street" is exactly the same. If each occurrence were not identical, we might perform a word frequency count on a document with invalid results because "street," "Street," "St.," and so on, would all be counted separately. Depending on the eventual use of this information, such problems could prove disastrous.

```
data _NULL_;
input address $50.;
text = PRXCHANGE('s/\s+([sS]t(reet)?|st\.)\s+/ St. /o',-1,address);
put text;

datalines;
103 Pennsylvania Ave NW, Washington, DC 20216
508 First St NW, Washington, DC 20001
650 1st St NE, Washington, DC 20002
3000 K Street NW, Washington, DC 20007
;
run;
```

Output 2.3: Log with Updated Data

```
103 Pennsylvania Ave NW, Washington, DC 20216
508 First St. NW, Washington, DC 20001
650 1st St. NE, Washington, DC 20002
3000 K St. NW, Washington, DC 20007
```

Note: There are often a number of ways to achieve the same outcome. Understanding the context of an application will help you determine the best RegEx pattern to use.

Example 2.4: Using the Capture Buffer

Revisiting Example 2.3, suppose we now want to also make the addresses available to a system that accepts only comma-separated values (CSV) files. This is a great opportunity to use the capture buffer. With only a couple of minor code changes, we can now process the data lines to be CSV ready.

The new line of code uses PRXCHANGE with a more complex RegEx that chunks the address into the street, city, state, and ZIP code components. And we see that the new line takes the previous output variable Text as the input argument, instead of `address`. Doing this allows us to make changes to the already changed text. If we were to use address, we would merely update the original data lines rather than building on the prior step.

In reviewing the parentheses elements in the new RegEx, we can see how the four address components are identified. On that same line, each of the four elements is placed via the buffer reference, with a comma and space immediately following. Reviewing Output 2.4, we see that the code produces the expected outcome.

```
data _NULL_;
input address $50.;
text = PRXCHANGE('s/\s+(Street|street|St|st|st\.)\s+/ St. /o',-1,address);
text2 = PRXCHANGE('s/(.+?),*?\s+?(\w+?),*?\s+?(\w+?)\s+?(\d+?)/$1, $2, $3,
$4/o',-1,text);
put text2;

datalines;
103 Pennsylvania Ave NW, Washington, DC 20216
508 First St NW, Washington, DC 20001
650 1st St NE, Washington, DC 20002
3000 K Street NW, Washington, DC 20007
;
run;
```

Output 2.4: Corrected Data in the Log

```
103 Pennsylvania Ave NW, Washington, DC, 20216
508 First St. NW, Washington, DC, 20001
650 1st St. NE, Washington, DC, 20002
3000 K St. NW, Washington, DC, 20007
```

So, what else could we do to the text? A number of things remain to be performed in order to make these addresses ready for advanced applications. For instance, "Ave" and "1st" should also be standardized. Building on the example code above is the fastest way to explore the options and become more comfortable with some of these concepts.

2.2.4 PRXPOSN

Description

PRXPOSN returns the matched information from specified capture buffers. This RegEx function requires the RXSUBSTR, PRXMATCH, PRXNEXT, or PRXCHANGE functions to be running before being used so that the capture buffer can be referenced. Also, RegEx_ID is required rather than the actual RegEx. Otherwise, PRXPOSN will not work—necessitating the use of PRXPARSE. The N input argument is numeric and refers to the capture buffer (without $).

Syntax

```
Text = PRXPOSN(RegEx_ID, N, Source_Text)
```

RegEx_ID: Unique RegEx identifier returned by PRXPARSE (input argument, required)

N: Integer value of the capture buffer (input argument, required)

Source_Text: The text variable or literal to be operated upon (input argument, required)

Text: Character variable assignment of captured text (output, required)

When we know the exact number of existing capture buffer elements (i.e., N is known), then we can use PRXPOSN without an issue. However, what happens when the number of elements is different from what we expect? If there are values in the capture buffer but we make a reference that is larger than those available (maybe there are three, but we make a reference to number 5), then a missing value is returned. However, if we reference capture buffer position 0 (N=0), then the entire pattern match is returned regardless of the buffer length.

The next function, PRXPAREN, is very helpful in creating robust code when you are using the capture buffers in conjunction with PRXPOSN. It is also important to write robust RegEx patterns to ensure that you prevent issues from popping up.

Example 2.5: Extracting Data with Capture Buffers

In order to make both the capture buffer concept and this new function more clear, we're going to walk through a concrete example. Suppose we want to process addresses for which the structure is well known and store various pieces in a SAS data set for later use. Since we know the layout of the address, the capture buffer arrangement and the application of PRXPOSN are both very straightforward.

First, we create the RegEx_ID variable Text by using the PRXPARSE function. Then, we perform a logical test using the PRXMATCH function in the IF statement. Notice that this is an implicit test of a match existing (no equal sign is used). If a match of the RegEx exists within the identified text source, then we assign the various capture buffer values to variables by using PRXPOSN (city, state, and zip).

Output 2.5 displays the values of `extract`, the data set created in our DATA step. As expected, we extracted the city, state, and ZIP code from each `datalines` entry. Later, we're going to build on this code to create a more sophisticated address extractor that includes the street information as well as the ability to include 9-digit zips.

```
data extract;
input address $50.;
text = PRXPARSE('/\s+(\w+),\s+(\w+)\s+(\d+)/o');
```

```
    if PRXMATCH(text, address) then
        do;
            city = PRXPOSN(text, 1, address);
            state = PRXPOSN(text, 2, address);
            zip = PRXPOSN(text, 3, address);
            output;
        end;
keep city state zip;
datalines;
103 Pennsylvania Ave NW, Washington, DC 20216
506 First St NW, Washington, DC 20001
650 1st St NE, Washington, DC 20002
3000 K Street NW, Washington, DC 20007
3000 K Street NW, Washington, DC, 20007
;
run;
proc print data=extract;
run;
```

Output 2.5: PROC PRINT Results

The SAS System

Obs	city	state	zip
1	Washington	DC	20216
2	Washington	DC	20001
3	Washington	DC	20002
4	Washington	DC	20007

2.2.5 PRXPAREN

Description

This function returns the numerical reference value of the largest capture buffer that contains data. It is therefore implicitly required that PRXSUBSTR, PRXMATCH, PRXNEXT, or PRXCHANGE be run prior to this function being used—just like with PRXPOSN. However, the only input argument is the RegEx_ID. Simply providing the RegEx is not an option, so this function must be used in conjunction with the PRXPARSE function.

Syntax

```
Paren=PRXPAREN(RegEx_ID)
```

RegEx_ID: Unique RegEx identifier returned by PRXPARSE (input argument, required)

Paren: Numerical reference value of the largest capture buffer (output, required)

Note: Since this function requires a RegEx_ID in lieu of the actual RegEx, it is implied that all precedents are then forced to use RegEx_ID instead of the RegEx as well—otherwise, PRXPAREN cannot be used.

What are we really trying to achieve with this function? Since it provides the length of the capture buffer by telling us the largest buffer position to contain text, we know exactly how many possible buffer values we can access. Because we know this, we can avoid errors when referencing them in code. It is worth noting that effective RegEx coding avoids many potential problems. However, it is always best practice to create fail-safe measures. In addition, we can use this function to identify which of several options has been triggered inside the source text.

Example 2.6: Identifying Capture Buffers

Ideally, whenever we want to use the PRXPOSN function, the data that we expect to be available in the source is available. However, we know that in reality, that is not always the case. So, we have to write code that can account for a reasonable amount of variability in any data that we might need to process. We are going to explore an advanced example later in this chapter that leverages the basic concepts outlined by this example.

Now, suppose we have a pattern with multiple possible matches embedded in it. How do we know which option allowed the pattern to create a match? In the code below, we see that it is possible to use PRXPAREN to answer this question. We have a simple pattern with three possible matches: "Dog", "Rat", and "Cat". Each is encapsulated by parentheses to create a capture buffer location. However, notice that the entire group is inside yet another set of parentheses. While unnecessary for practical purposes, this was done to demonstrate how capture buffers are numbered. Also note that this is not the most efficient way to write such code. We have sacrificed efficiency here in order to clarify how the buffers work. Notice in our output that "Dog" has a capture buffer of 2 despite being the first item in the OR list. Why? Because the outer set of parentheses is encountered first by SAS, thus creating a capture buffer element at position 1.

If we were to use PRXPOSN under each IF statement with position 1 in our argument list, we would see that each of the three cases below would be provided as output (i.e., when "Dog" is true, "Dog" would be in buffer 1 as well as in buffer 2, and so on). See Output 2.6 for the results.

```
data _null_;
   RegEx_ID=prxparse('/\b((Dog)|(Rat)|(Cat))\b/o');

   position=prxmatch(RegEx_ID, 'The Cat in the Hat');
   if position then paren=prxparen(RegEx_ID);
        put 'I matched capture buffer ' paren;

   position=prxmatch(RegEx_ID, 'The Rat in the Hat');
   if position then paren=prxparen(RegEx_ID);
        put 'I matched capture buffer ' paren;

   position=prxmatch(RegEx_ID, 'The Dog on the Roof');
   if position then paren=prxparen(RegEx_ID);
        put 'I matched capture buffer ' paren;
run;
```

As I have indicated, our goal here is to identify which particular capture buffer is used. This allows us to build more sophisticated functionality in the future, such as conditional information capture or standardization.

Output 2.6: Log Output

```
I matched capture buffer 4
I matched capture buffer 3
I matched capture buffer 2
```

2.3 Built-in SAS Call Routines

In this section, you learn about the PRX call routines available in SAS for performing many of the same RegEx tasks as the functions previously discussed, as well as some new ones. However, just like with all other call routines, PRX call routines cannot be used in expressions or assignment statements. The way they are implemented, and their ultimate functionality, is slightly different when compared to the functions. These differences are explored more thoroughly in the associated examples.

2.3.1 CALL PRXCHANGE

Description

This call routine performs the match-and-replace operation similar to that of the PRXCHANGE function. However, unlike the function version, the call routine must receive a RegEx identifier, without the option of using the associated RegEx instead. Also, there are some additional routine arguments not available in the function (result_length and truncation_value). The only required arguments are: RegEx_ID, Num_Times, and Input_string. All remaining arguments are optional.

Syntax

```
CALL PRXCHANGE(RegEx_ID, Num_Times, Input_string, Output_string, result_length,
trunc_value, num_changes)
```

RegEx_ID: Unique RegEx identifier returned by PRXPARSE (input argument, required)

Num_Times: Number of times the change is to be applied (input argument, required)

Input_string: Input text variable (input argument, required)

Output_string: Output text variable (input argument, optional). Default is Input_string.

result_length: Length of the characters put into Output_string (returned value, optional)

Trunc_value: Binary integer (0 or 1 only) value (returned value, optional). 1 means that the inserted text is longer than the text replaced. 0 means that the inserted text is either the same length or shorter than the text being replaced.

num_changes: The number of times the changes were made (returned value, optional)

Using the call routine in lieu of the function can often be cleaner from a coding perspective, especially when managing large programs. But there is a more practical reason for using this call routine instead of the function: accessing the additional functionality provided by the optional arguments. Since we have the ability to write changes directly back to the original variable, we can avoid creating new variables unnecessarily. This is especially useful when applying multiple data standardization filters to source text.

Note: Writing changes back to the existing variable makes them irreversible in the event of a mistake. So, while our ultimate use of this functionality requires the overwriting approach for sound memory management, creating new variables or data sets is ideal when you are still learning. It allows you to experiment and make mistakes without fear of making permanent changes to source data.

Example 2.7: Transforming Data

Let's look at basic usage for making changes to our source text. This is a simple example of how to use the call routine. Notice how compact this makes our code while maintaining functionality. Output 2.7 demonstrates that we have the anticipated functionality (replacing various forms of "street" with "St.").

```
data _NULL_;
input address $50.;
mypattern = PRXPARSE('s/\s+(Street|street|St|st|st\.)\s+/ St. /o');
CALL PRXCHANGE(mypattern,-1,address);
put address;

datalines;
103 Pennsylvania Ave NW, Washington, DC 20216
508 First St NW, Washington, DC 20001
650 1st St NE, Washington DC 20002
3000 K Street NW, Washington, DC 20007
;
run;
```

Output 2.7: Results in the SAS Log

```
103 Pennsylvania Ave NW, Washington, DC 20216
508 First St. NW, Washington, DC 20001
650 1st St. NE, Washington DC 20002
3000 K St. NW, Washington, DC 20007
```

Now that we've looked at a basic implementation of CALL PRXCHANGE, let's explore the optional arguments.

Example 2.8: Redacting Sensitive Data

In this example, we focus on developing your understanding of the optional elements in CALL PRXCHANGE. As a change of pace, we're going to develop a basic way to redact sensitive information. This is a frequent need, especially in the medical field, for protecting Personally Identifiable Information (PII).[1] Here, we're not going to eliminate all PII from the provided data because we are just demonstrating the functionality of CALL PRXCHANGE. However, this process is done more rigorously in Section 2.4 and on a larger scale.

In the code below, we start by creating a data set to pass into the DATA step, called `example`. This data set contains name, address, and phone number information in various configurations. In the DATA step, we create a RegEx_ID using PRXPARSE, and then use CALL PRXCHANGE to execute the changes prescribed by the RegEx. Notice that one RegEx_ID is commented out. The RegEx in that line behaves very differently from the initial RegEx_ID definition, which allows us to demonstrate the trunc_val and num_changes options. We use this commented RegEx_ID to create Output 2.8. After the DATA step, we perform a PROC PRINT to create the output shown in both Output 2.8 and Output 2.9.

```
data example;
   input text $80.;
   datalines;
Ken can be reached at (801)443-9876
103 Pennsylvania Ave NW, Washington, DC 20216
JP's address is:
650 1st St NE, Washington DC 20002
Carla's information is: (910)998-8762
```

```
3000 K Street NW, Washington, DC 20007
Eric can be reached at: (321) 456-7890
508 First St NW, Washington, DC 20001
;
run;

data changed;
   set example;

   *RegEx_ID = PRXPARSE('s/\d+/***NUMBER REMOVED***/o');
   RegEx_ID = PRXPARSE('s/\([1-9]\d\d\)\s?[1-9]\d\d-\d\d\d\d/*REDACTED*/o');
   Call PRXCHANGE(RegEx_ID, -1, text, text, length, trunc_val, num_changes);
   put text=;
run;

proc print data=changed;
run;
```

Simple Insert Results

As we can see in the results below, the phone numbers have been redacted in the original text by using the value *REDACTED*. The rest of the data set shows our optional variable values. Length is the total length of the string written to Text (we just wrote back to the old string this time). trunc_val is 0 for every row because the inserted value is no longer than the original phone numbers. In fact, lines 1, 5, and 7 shrink because the inserted content is shorter. And finally, num_changes records the number of times the phone numbers were redacted on each line (multiple phone numbers per line would have resulted in that number occurring in this column).

Output 2.8: SAS PROC PRINT Results

	The SAS System				
Obs	text	RegEx_ID	length	trunc_val	num_changes
1	Ken can be reached at *REDACTED*	1	77	0	1
2	103 Pennsylvania Ave NW, Washington, DC 20216	1	80	0	0
3	JP's address is:	1	80	0	0
4	650 1st St NE, Washington DC 20002	1	80	0	0
5	Carla's information is: *REDACTED*	1	77	0	1
6	3000 K Street NW, Washington, DC 20007	1	80	0	0
7	Eric can be reached at: *REDACTED*	1	76	0	1
8	508 First St NW, Washington, DC 20001	1	80	0	0

More Advanced Insert Results

The commented RegEx_ID definition creates very different output for Output 2.9—a longer replacement value that occurs for every group of numbers (***NUMBER REMOVED*** is inserted). The variables are all the same as in Output 2.8, but notice how the values change. For instance, trunc_val now equals 1 every time a redaction occurred, and num_changes is frequently greater than 1. Also, notice something else very important about this output: some lines of text are actually truncated!

Remember, the trunc_val variable being set to 1 does not mean that data loss is certainly going to occur. Instead, it means that it could occur. Think of this as a warning flag telling us, "Hey, keep a look out for a problem." And a problem is what we would indeed have for some of these lines of text. The insertion of longer text pushes all following text to the right (beyond the 80-character length defined for the variable Text). Now, when there is a significant amount of white space to the right of our text, this doesn't result in an issue. However, when there is valuable information to the right of our inserted text, we will likely have data loss. Regardless how small the loss of data, the integrity of our entire data set is compromised if we do not design code that avoids this problem. We discuss this concept a bit more in Section 2.4.

Output 2.9: SAS PROC PRINT Results

	The SAS System				
Obs	text	regex_id	length	trunc_val	num_changes
1	Ken can be reached at (***NUMBER REMOVED***)***NUMBER REMOVED***-***NUMBER REMOV	1	80	1	3
2	***NUMBER REMOVED*** Pennsylvania Ave NW, Washington, DC ***NUMBER REMOVED***	1	80	1	2
3	JP's address is:	1	80	0	0
4	***NUMBER REMOVED*** ***NUMBER REMOVED***st St NE, Washington DC ***NUMBER REMOV	1	80	1	3
5	Carla's information is: (***NUMBER REMOVED***)***NUMBER REMOVED***-***NUMBER REM	1	80	1	3
6	***NUMBER REMOVED*** K Street NW, Washington, DC ***NUMBER REMOVED***	1	80	1	2
7	Eric can be reached at: (***NUMBER REMOVED***) ***NUMBER REMOVED***-***NUMBER RE	1	80	1	3
8	***NUMBER REMOVED*** First St NW, Washington, DC ***NUMBER REMOVED***	1	80	1	2

2.3.2 CALL PRXPOSN

Description

This call routine takes the RegEx_ID provided by PRXPARSE and the numerical capture buffer position N as inputs. It produces the matching Position and Length as outputs.

Syntax

```
CALL PRXPOSN(RegEx_ID, N, Position, Length)
```

RegEx_ID: Unique RegEx identifier returned by PRXPARSE (input argument, required)

N: Integer value of the capture buffer (input argument, required)

Position: Integer value of the character position for the first character in the matched pattern (returned value, required)

Length: Integer value for the length of the matched pattern (returned value, optional)

This call routine takes in the RegEx_ID (RegEx is not allowed!) and capture buffer, and returns the exact locations where it occurs in the most recent match. The match results from PRXMATCH, PRXCHANGE, PRXSUBSTR, or CALL PRXNEXT (discussed in Section 2.3.4) must exist in order for CALL PRXPOSN to work properly. We then must use the SUBSTR function to extract the identified text.

Example 2.9: Context-specific Algorithm Development

Sometimes it's useful to condition code behavior on specific words occurring in text. In this example, you'll see how the functionality of CALL PRXPOSN can be used in combination with PRXPAREN, PRXMATCH, PRXPARSE, and SUBSTR to do just that.

First, we create the RegEx_ID by using PRXPARSE, which is then passed to PRXMATCH. If a result from PRXMATCH exists (i.e., a pattern match is found), then we determine which of the capture buffers in the pattern is matched via PRXPAREN. The output of PRXPAREN is used as input to the CALL PRXPOSN routine to create the Position and Length outputs. SUBSTR is then used to extract the specified text. We then print a message, depending on the buffer position. See the results in Output 2.10.

```
data _null_;
input text $50.;
    RegEx_ID=prxparse('/((Dog)|(Rat)|(Cat))/o');

    if prxmatch(RegEx_ID, text) then do;
            paren=prxparen(RegEx_ID);
            CALL PRXPOSN(RegEx_ID, paren, position, length);
            buffer = substr(text, position, length);
            put 'I matched capture buffer ' paren 'with ' buffer;
            end;

if paren=2 then put 'I love dogs!';
else put 'I cannot stand a ' buffer'!';

datalines;
The Cat in the Hat
The Rat in the Hat
The Dog on the Roof
;
run;
```

I added the commentary about cats, rats, and dogs to show a second way to perform conditioning on the parsed text. Obviously, you could perform more interesting things, and it should be fun to experiment with in the future. We use this concept in later chapters to build out some interesting functionality.

Output 2.10: Log Output of Code Behavior

```
I matched capture buffer 4 with Cat
I cannot stand a Cat !
I matched capture buffer 3 with Rat
I cannot stand a Rat !
I matched capture buffer 2 with Dog
I love dogs!
```

2.3.3 CALL PRXSUBSTR

Description

This call routine takes RegEx_ID and Source_Text as inputs, and returns Position and Length as outputs. Using the actual RegEx is not an option.

Syntax

```
CALL PRXSUBSTR(RegEx_ID, Source_Text, Position, Length)
```

RegEx_ID: Unique RegEx identifier returned by PRXPARSE (input argument, required)

Source_Text: The text to be operated upon (input argument, required)

Position: Integer value of the character position for the first character in the matched pattern (returned value, required)

Length: Integer value for the length of the matched pattern (returned value, optional)

This call routine is used extensively in Information Extraction applications like those discussed in Chapter 4. Since only the position and length of matches are identified by CALL PRXSUBTR, it must be used in conjunction with a function like SUBSTR in order to extract the actual text.

Example 2.10: Information Extraction

In this example, we revisit the now-familiar example code from Chapter 1. It is a great example of how you can leverage the CALL PRXSUBSTR in many applications.

The code below creates a RegEx pattern to search for all occurrences of "Smith" in our source text. It then generates a RegEx_ID using PRXPARSE. The code then uses CALL PRXSUBSTR to search through source text with the provided pattern and return the position and length of matching text. As you know by now, this could have been a much more complex pattern, but the simplicity here helps to highlight the functionality that we are focused on learning. After the call routine, the code checks to see whether the position variable (Position) is 0, which is the default value indicating that it did not find a match. If a position does exist, the code proceeds to use SUBSTR to capture text from the source using the position and length obtained by CALL PRXSUBSTR. Results are then output to the log. See Output 2.11.

```
data _NULL_;
pattern = "/Smith/o"; /*<--Edit the pattern here.*/
pattern_ID = PRXPARSE(pattern);
input some_data $50.;
CALL PRXSUBSTR(pattern_ID, some_data, position, length);
if position ^= 0 then
   do;
      match = substr(some_data, position, length);
      put match:$QUOTE. "found in " some_data:$QUOTE.;
   end;
datalines;
Smith, BOB A.
ROBERT Allen Smith
Smithe, Cindy
103 Pennsylvania Ave. NW, Washington, DC 20216
508 First St. NW, Washington, DC 20001
650 1st St NE, Washington, DC 20002
3000 K Street NW, Washington, DC 20007
1560 Wilson Blvd, Arlington, VA 22209
1-800-123-4567
1(800) 789-1234
;
run;
```

As we expected, the output shows the various occurrences of "Smith" from within the provided data lines. This example brings us full circle with the above code, pulling all of the pieces together.

Output 2.11: Log Results for "Smith"

```
"Smith" found in "Smith, BOB A."
"Smith" found in "ROBERT Allen Smith"
"Smith" found in "Smithe, Cindy"
```

2.3.4 CALL PRXNEXT

Description

This routine searches through Source_Text, between the Start and Stop positions, for the pattern associated with RegEx_ID. It returns the Position and Length of the location.

Syntax

```
CALL PRXNEXT(RegEx_ID, Start, Stop, Source_Text, Position, Length)
```

RegEx_ID: Unique RegEx identifier returned by PRXPARSE (input argument, required)

Start: Numerical constant, variable, or expression containing the starting character position to begin the search (input argument, required)

Stop: Numerical constant, variable, or expression containing the last character position to use in the search. If the value is -1, the stop position becomes the last non-blank character position in the source. (input argument, required)

Source_Text: The text to be operated upon (input argument, required)

Position: Integer value of the character position for the first character in the matched pattern (returned value, required)

Length: Integer value for the length of the matched pattern (returned value, required)

This call routine can be used for two applications:

1. searching for a pattern within a defined range
2. searching for a pattern iteratively throughout text, including multiple occurrences per line

The first application of CALL PRXNEXT is a straightforward implementation of the routine's parameters. However, the second usage is less apparent from its definition. Therefore, we focus on that usage of the routine in our example.

Example 2.11: Pattern Matching Multiple Times per Line

Being able to identify a pattern any number of times in a particular line of text is valuable for many practical applications. For example, performing word frequency counts clearly requires this ability in order for accurate counts to be obtained.

The code below shows how to use CALL PRXNEXT to identify multiple occurrences of our pattern on each row from the data-lines source. The pattern is defined to match on any string that is three word characters (\w) in length and that ends with "un". The Start and Stop variables are initialized to character positions 1 and Length(some_data) respectively. These variables must be provided with initial values for the routine's first use. However, subsequent calls automatically reset the start position to the character position immediately following the most recent successful match. This is a fact that we take advantage of with the DO WHILE loop below. If we were to eliminate the loop portion of code, we would merely be searching for the pattern in a defined range (use the above application #1), but having the loop allows us to achieve the desired functionality (use the above application #2). See the results in Output 2.12.

```
data _NULL_;
input some_data $50.;

pattern = "/\wun/o";
pattern_ID = PRXPARSE(pattern);
start = 1;
stop = length(some_data);

CALL PRXNEXT(pattern_ID, start, stop, some_data, position, length);
   do while (position > 0);
       found = substr(some_data, position, length);
       put "Line:" _N_ found= position= length=;

       CALL PRXNEXT(pattern_ID, start, stop, some_data, position, length);
   end;

datalines;
Running Runners who run.
Runners who think running is fun.
"Fun Runs" are not-so-fun runs for me.
Let's run at the next reunion.
;
run;
```

Log Output of Pattern Match Results

Output 2.12 contains the literal string that was found, its position, and its length. As we should expect by now, the pattern that is created ignores the surrounding text—which makes the example slightly more interesting. Review the output (count the character locations in the data lines), and notice that we did indeed achieve the desire results.

Output 2.12: Log Output of Pattern Match Results

```
Line:1 found=Run position=1  length=3
Line:1 found=Run position=9  length=3
Line:1 found=run position=21 length=3
Line:2 found=Run position=1  length=3
Line:2 found=run position=19 length=3
Line:2 found=fun position=30 length=3
Line:3 found=Fun position=2  length=3
Line:3 found=Run position=6  length=3
Line:3 found=fun position=23 length=3
Line:3 found=run position=27 length=3
Line:4 found=run position=7  length=3
Line:4 found=eun position=24 length=3
```

2.3.5 CALL PRXDEBUG

Description

This routine is used to perform debugging of all PRX functions and call routines, and accepts only one input.

Syntax

```
CALL PRXDEBUG (ON-OFF)
```

> ON-OFF: Numerical constant, variable, or expression. If it equals 0, then debugging is turned off, but any positive value turns it on. (input argument, required)

This routine prints step-by-step output to the log, enabling a low-level understanding of any PRX program. However, be prepared—this routine can create voluminous output. It is best to use it in a targeted way at first in order to understand how a specific function or routine is working (or not working). If we were to use this routine for an entire program, we should be ready to read very large amounts of procedural output, which is an inefficient approach to diagnosing issues. It is best to perform gross-level diagnostics using PUT statements and dummy variables, thus narrowing the focus to a specific code segment before using CALL PRXDEBUG. In practice, this is the fastest approach to identifying the source of logical errors.

Example 2.12: Debugging the PRXPARSE Function

In keeping with our goal of using the CALL PRXDEBUG in a targeted way to debug code, we are going to apply it only to the PRXPARSE function in the code below. Notice that we have to turn it on and off at different points in the code in order to identify the segment to which we want our debug output limited. See the results in Output 2.13.

```
data _null_;
input text $50.;

CALL PRXDEBUG(1);
    RegEx_ID=prxparse('/((Dog)|(Rat)|(Cat))/o');
CALL PRXDEBUG(0);
    if prxmatch(RegEx_ID, text) then do;
        paren=prxparen(RegEx_ID);
        CALL PRXPOSN(RegEx_ID, paren, position, length);
        buffer = substr(text, position, length);
        put 'I matched capture buffer ' paren 'with ' buffer;
        end;

    if paren=2 then put 'I love dogs!';
    else put 'I cannot stand a ' buffer'!';

datalines;
The Cat in the Hat
The Rat in the Hat
The Dog on the Roof
;
run;
```

Debugging Information Printed to the Log

Reviewing the output in Output 2.13, we see that the debugging information for just a single PRX function can be quite large, thus reinforcing my earlier point about limiting the scope of CALL PRXDEBUG.

The first line denotes compilation of a RegEx within the PRXPARSE function. The next line shows us the compiled RegEx size and starting location for the lines that follow. Specifically, the size of 26 refers to the 26 lines of compiled RegEx code (numbers on the left with a trailing semi-colon), and first refers to the first line of code execution. The numbers in parentheses to the right of each line correspond to labels for the compiled RegEx (these labels work much like our pseudo code labels in Chapter 1).

Lines 1 through 26 are the compiled steps within our RegEx, and they become easy to follow once we understand what each represents. For instance, the various OPEN and CLOSE statements correspond to our opening and closing parentheses; BRANCH corresponds to the OR tests between the inner three parenthesis pairs; and EXACT is for the string literal match of the associated word. END obviously means the end of the subroutine.

The remaining output is just the rest of our code running as normal. Should our code be malfunctioning, we would not likely see such normal output when using CALL PRXDEBUG.

Output 2.13: Debugging Information Printed to the Log

```
❶ Compiling REx `((Dog)|(Rat)|(Cat))'
  size 26 first at 3
❷    1: OPEN1(3)
     3:    BRANCH(10)
     4:      OPEN2(6)
     6:        EXACT <Dog>(8)
     8:      CLOSE2(24)
    10:    BRANCH(17)
    11:      OPEN3(13)
    13:        EXACT <Rat>(15)
    15:      CLOSE3(24)
    17:    BRANCH(24)
    18:      OPEN4(20)
    20:        EXACT <Cat>(22)
    22:      CLOSE4(24)
    24: CLOSE1(26)
❸   26: END(0)
❹ minlen 3

  I matched capture buffer 4 with Cat
  I cannot stand a Cat !
  I matched capture buffer 3 with Rat
  I cannot stand a Rat !
  I matched capture buffer 2 with Dog
  I love dogs!
```

❶ The compilation process begins for the quoted RegEx contained by PRXPARSE.

❷ Notice that each OPEN and CLOSE pair have the same number (OPEN1 and CLOSE1). These numbers correspond to the numerical value of the capture buffer that was formed by that set of parentheses.

❸ Each line ends with a number enclosed in parentheses, denoting the next line to jump to from that line. However, the END tag shows a jump to 0, which takes us out of the subroutine.

❹ The minlen field defines the minimum length for the match to be 3. This information is used by subsequent functions and routines when using this compiled pattern.

Moving the placement of our debug routine call should prove to yield some interesting, and potentially rather long, output. Doing so is the best way to become more familiar with the low-level operations SAS is performing behind the scenes of our PRX code.

Significant amounts of information can be provided by the PRXDEBUG output, but a much deeper study of debug output is outside the scope of this text. For more information about debug output and its meaning, visit the SAS Support website.[2]

2.3.6 CALL PRXFREE

Description

This call routine releases memory resources associated with a RegEx, using its unique RegEx_ID. Subsequent references to this identifier return a missing value.

Syntax

```
CALL PRXFREE(RegEx_ID)
```

RegEx_ID: Unique RegEx identifier returned by PRXPARSE (input argument, required)

This routine is used to free up memory for a specified RegEx_ID and becomes very important for managing the memory of large programs. Remember, there is much more happening behind the scenes of the RegEx_ID construction, despite merely having a numerical identifier. (See CALL PRXDEBUG.) It can't be stressed enough that memory management can be a significant problem for large programs if not handled properly. Although SAS still handles memory cleanup to avoid memory leaks when a session ends, it is possible to run into memory limitations within a single session. Think very strategically about which RegEx_IDs—or any other variables for that matter—are necessary for each chunk of code.

Example 2.13: Releasing Memory with CALL PRXFREE

In order to demonstrate the functionality of CALL PRXFREE, we are revisiting a new version of the example code for PRXCHANGE. However, instead of printing output as in the original example, we are going to concern ourselves only with the results related to CALL PRXFREE. (See Output 2.14.)

As we can see in the code below, the PUT statement is used to print the values of Street_RXID and AddParse_RXID to the log for each run through the DATA step (creating four writes to the log). However, using the IF statement, we run the CALL PRXFREE routine on the last record to release the memory associated with both RegEx_IDs. Then, we print the results to the log. This creates a fifth write to the log, but the values are missing this time because our routine was successful at releasing the memory allocated for them—making them unrecoverable.

```
data sample;
input address $50.;
datalines;
103 Pennsylvania Ave NW, Washington, DC 20216
508 First St NW, Washington, DC 20001
650 1st St NE, Washington DC 20002
3000 K Street NW, Washington, DC 20007
;
run;
```

```
data _null_;
set sample end=last;
Street_RXID = PRXPARSE('s/\s+?(S|s)\w+?\s+/ St. /o');
AddParse_RXID = PRXPARSE('s/(.+?),*?\s+?(\w+?),*?\s+?(\w+?)\s+?(\d+?)/$1, $2, $3,
$4/o');
text = PRXCHANGE(Street_RXID,-1,address);
text2 = PRXCHANGE(AddParse_RXID,-1,text);
put Street_RXID AddParse_RXID;

if last then do;
    CALL PRXFREE(Street_RXID);
    CALL PRXFREE(AddParse_RXID);
    put Street_RXID AddParse_RXID;
    end;
run;
```

Output 2.14: Log Printout

```
1  2
1  2
1  2
1  2
.  .
```

The missing values displayed for each of the RegEx_ID variables demonstrate that the CALL PRXFREE routine released all memory associated with both.

2.4 Applications of RegEx

In this section, we explore some real-world applications of RegEx with SAS, demonstrating a wide variety of scenarios in which we can implement what you have learned thus far. The general categories that these examples have been placed under do not to imply that we are limited in what we can do (see Chapter 1), nor do they imply a lack of overlap between some of the examples.

In order to execute some of the applications to follow, we need a good source of addresses, phone numbers, names, birthdates, and Social Security numbers. Obviously, these sources are hard to come by for experimentation (that is a lot of personal information!). Therefore, for the sake of practice, I created some code to randomly generate these more sensitive data items. The code for these random Personally Identifiable Information (PII) elements is presented in Appendix A, and is not repeated for the different applications. Access to real data sources is always preferable when developing robust code, but the random generator does a fair job of inserting some commonly occurring variability into these elements (see Appendix A for documentation). Feel free to experiment with it to obtain a different number of records or greater variability in the information.

While all of the examples in this section are realistic, there is still room for improvement. However, we can do only so much in this book. So, at the end of each section, I assign you some homework—suggested assignments for how to improve the code already provided. These items should prove especially interesting for advanced programmers.

2.4.1 Data Cleansing and Standardization

As data sets go, the randomly generated data set that we are going to work with is fairly clean. The simple fact is we can't explore all of the ways that data can be dirty in the real world (this book would never end!). However, using some realistic data, we can test our ability to develop RegEx code to process and clean some common problems in such data sources. This exercise will prepare you to go out in the real world and tackle a wide variety of things that you might encounter because you will have the necessary tools in your toolbox.

So, let's start by reviewing the data elements that we need in order to clean and standardize, and what things we need to check for in such sources.

Firstname
  The person's first name. This piece of data should contain character values only.

Surname
  The person's last name. This piece of data should contain character values only.

PhoneNumber
  The person's phone number. Phone numbers in different countries are written very differently, so we must be prepared to properly parse a variety of formats—especially since it is so easy for a business contact to be from or located in a different country.

SSN
  The person's Social Security number. We should see only segments of numbers separated by dashes or spaces, and we need to enforce this formatting.

DOB
  The person's date of birth. This can be represented in a few ways, but we primarily see the classic 8-digit format in the US. European dates represent the day before the month. This is where context is very important, because it is difficult to detect this format unless the obvious value thresholds are crossed.

Address
  The person's address. This data has the most natural variability and is the most interesting to parse. We primarily need to be concerned with abbreviations, punctuation, and ZIP code lengths.

Now, as I hinted previously, data cleansing and standardization is accomplished by creating what amount to filters. These filters are a series of RegEx functions and routines applied in succession so as to yield incremental changes as each one is applied. Implemented in the correct order, we can clean up some very messy data for later use. Fortunately for us, the data set created by the PII generator (see Appendix A) is relatively tame. But only a few things need to change in order for it to become scary data. Regardless, there are a few things that we must fix in order to make use of the entire data set in its current state. For example, when we look at observation 16 from the data set (Output 2.15), we see a few issues with the address.

Output 2.15: PII Observation 16

| 16 | MATTHEW | MARTIN | +2.623.941.6074 | 436-07-9380 | 10/29/1975 | 10th St. & Constitution Ave. NW, Washington, DC 20560 |

In addition to having decimals immediately following the abbreviations for street and avenue, we see that the & symbol is used. Both these issues will become problematic when we attempt to parse the address into street, city, state, and zip.

Developing what needs to be fixed in any data set often can't be done blindly. There are basic things that we can apply to any data source, such as trimming excess spaces, and so on. However, it is advisable to pull samples of data in order to understand its quality issues before you develop RegEx patterns for cleaning.

In the code below, I created a series of cleansing and standardization steps using PRXPARSE, PRXMATCH, CALL PRXCHANGE, and PRXPOSN. Notice how clean our code is by using CALL PRXCHANGE in lieu of the function version.

```
data CleanPII;
set PII;

ChangeAND = PRXPARSE('s/\x26/and/o');              ❶
ChangeSTR = PRXPARSE('s/\s(St\.|St)/ Street/o');
ChangeAVE = PRXPARSE('s/\s(Ave\.|Ave)/ Avenue/o');
ChangeRD  = PRXPARSE('s/\s(Rd\.|Rd)/ Road/o');
ChangeDASH = PRXPARSE('s/\s*(\.|\(|\))\s*/-/o');
ChangePLUS = PRXPARSE('s/\+//o');
ChangeSPAC = PRXPARSE('s/ //o');

/*Cleaning Address*/                               ❷
CALL PRXCHANGE(ChangeAND,-1,address);
CALL PRXCHANGE(ChangeSTR,-1,address);
CALL PRXCHANGE(ChangeAVE,-1,address);
CALL PRXCHANGE(ChangeRD ,-1,address);

/*Cleaning Phone Number*/
CALL PRXCHANGE(ChangeDASH,-1,PhoneNumber);
CALL PRXCHANGE(ChangePLUS,-1,PhoneNumber);
CALL PRXCHANGE(ChangeSPAC,-1,PhoneNumber);

/*Cleaning SSN*/
CALL PRXCHANGE(ChangeDASH,-1,SSN);
CALL PRXCHANGE(ChangeSPAC,-1,SSN);

/*Cleaning DOB*/
CALL PRXCHANGE(ChangeSPAC,-1,DOB);

drop ChangeAND ChangeSTR ChangeAVE ChangeRD ChangeDASH ChangePLUS ChangeSPAC;
run;

data FinalPII;
set CleanPII;

/*Parsing the address into its discrete parts*/
Addr_Pattern =
PRXPARSE('/^(\w+(\s\w+)*\s\w+),\s+(\w+\s*\w+),\s+(\w+)\s+((\d{5}\s*?-
\s*?\d{4})|\d{5})/o');
   if PRXMATCH(Addr_Pattern, address) then       ❸
     do;
   Street = PRXPOSN(Addr_Pattern, 1, address);
       City = PRXPOSN(Addr_Pattern, 3, address);
        State = PRXPOSN(Addr_Pattern, 4, address);
```

```
        Zip = PRXPOSN(Addr_Pattern, 5, address);
    end;
drop Addr_Pattern address;
run;

proc print data=finalpii;
run;
```

❶ First, we create a series of replacement RegEx pattern identifiers using PRXPARSE. We maintain the "change" naming convention to denote that each identifier represents a RegEx pattern for changing the source.

❷ Here we apply specific RegEx_ID's by using the CALL PRXCHANGE routine to clean each of the variables in different ways.

❸ Finally, we parse the original address data field into its constituent parts: street, city, state, and zip. The PRXPOSN function grabs each piece of the address by identifying the associated capture buffer. Notice that we have to skip buffer location 2 because the second bracket set is used for logical separation inside the first bracket set (which creates buffer location 1). Referencing buffer location 2 would provide only a subset of the street information that we need.

As we can see in Output 2.16, our resulting data set now has clean, standardized data in each field. Such data makes future analysis and manipulation much easier and more accurate. This exercise should serve as a nice warm-up for many other such applications.

Output 2.16: Cleaned and Standardized PII Data Set

Obs	Firstname	Surname	PhoneNumber	SSN	DOB	Street	City	State	Zip
1	JAMES	SMITH	2-746-475-4589	539-71-9216	12/17/1986	1776 D Street NW	Washington	DC	20006
2	JOHN	JOHNSON	0-439-270-9250	189-03-1020	3/25/1981	1600 Pennsylvania Avenue NW	Washington	DC	20500
3	WILLIAM	WILLIAMS	6-281-794-3626	971-45-0631	10/2/1986	600 14th Street NW	Washington	DC	20005
4	DAVID	JONES	9-349-208-5935	277-05-1098	8/9/1985	1321 Pennsylvania Avenue NW	Washington	DC	20004
5	THOMAS	BROWN	3-287-870-2874	123-22-9494	1/26/1980	2470 Rayburn Hob	Washington	DC	20515
6	CHRISTOPHER	DAVIS	6-639-100-7721	688-49-1392	5/20/1992	101 Independence Avenue SE	Washington	DC	20540
7	DANIEL	MILLER	8-277-323-9564	675-36-2461	9/9/1977	511 10th Street NW	Washington	DC	20004
8	PAUL	WILSON	6-863-034-6857	304-59-8869	5/18/1980	450 7th Street NW	Washington	DC	20004
9	MARK	MOORE	5-697-801-1886	102-01-9574	7/29/1991	2 15th Street NW	Washington	DC	20007
10	DONALD	TAYLOR	3-019-416-1550	439-72-6850	2/1/1982	3700 O Street NW	Washington	DC	20057
11	GEORGE	ANDERSON	0-036-200-7891	116-12-3837	8/10/1983	3001 Connecticut Avenue NW	Washington	DC	20008
12	KENNETH	THOMAS	1-705-466-8443	549-54-0433	5/22/1976	3101 Wisconsin Avenue NW	Washington	DC	20016
13	EDWARD	JACKSON	7-841-664-8908	461-17-1160	9/6/1977	800 Florida Avenue NE	Washington	DC	20002
14	ANTHONY	WHITE	8-451-939-9401	374-43-5208	5/7/1979	1 First Street NE	Washington	DC	20543
15	KEVIN	HARRIS	7-690-620-4418	877-45-2254	6/10/1984	600 Independence Avenue SW	Washington	DC	20560
16	MATTHEW	MARTIN	2-623-941-6074	436-07-9380	10/29/1975	10th Street and Constitution Avenue NW	Washington	DC	20560
17	LINDA	THOMPSON	6-564-897-1662	500-98-4809	3/10/1983	555 Pennsylvania Avenue NW	Washington	DC	20001
18	MARGARET	GARCIA	5-894-411-7166	772-03-0744	10/16/1992	1000 5th Avenue	New York	NY	10028
19	LISA	MARTINEZ	1-167-511-7529	426-25-2712	8/18/1978	64th Street and 5th Avenue	New York	NY	10021
20	KAREN	ROBINSON	5-395-540-6931	761-09-3862	10/18/1974	10 Lincoln Center Plaza	New York	NY	10023
21	DONNA	CLARK	7-781-276-1544	131-47-7120	11/3/1991	Pier 86 W 46th Street and 12th Avenue	New York	NY	10036
22	CAROL	RODRIGUEZ	9-737-785-2677	280-95-9464	12/28/1985	350 5th Avenue	New York	NY	10118
23	SHARON	LEWIS	0-918-040-2361	896-33-5968	1/31/1975	405 Lexington Avenue	New York	NY	10174
24	MICHELLE	LEE	9-488-783-8608	735-62-9285	10/3/1984	1 Albany Street	New York	NY	10006
25	KIMBERLY	WALKER	7-823-096-2389	109-02-1649	9/17/1990	1585 Broadway	New York	NY	10036

Homework

1. Include more standard abbreviations than the few we currently have (e.g., Parkway, Court, and so on).
2. Standardize two-digit years in the date field.
3. Create a method of handling state abbreviations with decimals.
4. Use CALL PRXFREE to clean up the RegEx_IDs used in the code.
5. Enhance the data set with a Census tract lookup using the address fields.
6. Enhance the RegEx to handle multiple spaces between words, spaces before commas, and punctuation in unexpected places.

2.4.2 Information Extraction

Parsing large volumes of text to generate structured data sets is a common, valuable use of RegEx capabilities. For example, we might want to collect information from a technology blog or website that contains valuable customer feedback about our product. Such information could not easily or cheaply be gathered by hand for the sake of further analysis. Due to the wide variety of possible sources from which we might need to extract information, as well as the wide array of end goals for the information, there are a number of approaches to accomplishing this task. Given the proliferation of tag-based languages such as HTML and XML, we need to be prepared to effectively extract information from them.

Now, due to the semi-structured nature of tag-based languages, they are certainly easier to process than one might anticipate. All such languages have paired opening and closing tags for defining various pieces of information. We can leverage this fact to properly dissect them and extract the information that we need.

With more sophisticated techniques at our disposal (like using the SAS macro facility), we could actually "learn" the embedded data elements and extract the data associated with the discovered variables. However, emphasis on these techniques is beyond the scope of this book. Therefore, we need to know the various tags that we are looking for in the XML or HTML source in advance. We will use this approach to process the XML file in our example.

Going back to the SEC administrative proceedings example from Chapter 1, let's parse and extract the information from the associated sample file.

Figure 2.1: SEC XML Sample[3]

```xml
<?xml version="1.0" encoding="ISO-8859-1"?>
- <root>
   - <administrative_proceeding>
        <url>http://www.sec.gov/litigation/admin/2009/34-61262.pdf</url>
        <release_number>34-61262</release_number>
        <release_date>Dec. 30, 2009</release_date>
        <respondents>Stephen C. Gingrich</respondents>
     </administrative_proceeding>
   - <administrative_proceeding>
        <url>http://www.sec.gov/litigation/admin/2009/34-61256.pdf</url>
        <release_number>34-61256</release_number>
        <release_date>Dec. 30, 2009</release_date>
        <respondents>Gabelli Funds LLC</respondents>
     </administrative_proceeding>
   - <administrative_proceeding>
        <url>http://www.sec.gov/litigation/admin/2009/34-61255.pdf</url>
        <release_number>34-61255</release_number>
        <release_date>Dec. 30, 2009</release_date>
        <respondents>Gabelli Funds LLC</respondents>
     </administrative_proceeding>
   - <administrative_proceeding>
        <url>http://www.sec.gov/litigation/admin/2009/34-61252.pdf</url>
        <release_number>34-61252</release_number>
        <release_date>Dec. 29, 2009</release_date>
        <respondents>Banc One Investment Advisors Corporation and Mark A. Beeson</respondents>
     </administrative_proceeding>
```

The primary concern is to effectively extract information from within the known XML tags that contain data, namely: url, release_number, release_date, and respondents. As you can see in the figure above, there are some other XML tags in the document, but they aren't relevant to the task at hand. For instance, root and administrative_proceeding don't contain data independent of the previously mentioned tags. They merely serve administrative functions in the context of XML for properly organizing the information for consumption by a system that reads XML directly.

```
data SECFilings;
infile 'F:\Unstructured Data
Analysis\Chapter_2_Example_Source\administrative_proceedings_2009.xml'
length=linelen lrecl=500 pad;
varlen=linelen-0;
input source_text $varying500. varlen; ❶
```

```
format ReleaseNumber $20. ReleaseDate $20. Respondents $500. URL $500.;
start = 1;
stop = varlen;
Pattern_ID = PRXPARSE('/\<(\w+)\>(.+?)\<\/(\w+)\>/o'); ❷
CALL PRXNEXT(pattern_ID, start, stop, source_text, position, length); ❸
   DO WHILE (position > 0);
   tag = PRXPOSN(pattern_ID,1,source_text);
   if tag='url' then URL = PRXPOSN(pattern_ID,2,source_text);
   else if tag='release_number' then ReleaseNumber =
PRXPOSN(pattern_ID,2,source_text);
   else if tag='release_date' then ReleaseDate =
PRXPOSN(pattern_ID,2,source_text);
   else if tag='respondents' then do;
           Respondents = PRXPOSN(pattern_ID,2,source_text); ❹
           put releasenumber releasedate respondents url;
           output;
           end;
      retain URL ReleaseNumber ReleaseDate Respondents; ❺
      CALL PRXNEXT(pattern_ID, start, stop, source_text, position, length);
   end;
keep ReleaseNumber ReleaseDate Respondents URL;
run;
proc print data=secfilings;
run;
```

❶ We begin by bringing data in from our XML file source via the INFILE statement, using the length, LRECL, and pad options. (See the SAS documentation for additional information about these options.) Next, using the INPUT statement, the data at positions 1-varlen in the Program Data Vector (PDV) are assigned to source_text. Because we set LRECL=500, we cannot capture more than 500 bytes at one time, but we can capture less. For this reason, we use the format $varying500.

❷ Using the PRXPARSE function, we create a RegEx pattern identifier, Pattern_ID. This RegEx matches on a pattern that starts with an opening XML tag, contains any number and variety of characters in the middle, and ends with a closing XML tag.

❸ Just like our example in Section 2.3.4, we make an initial call the CALL PRXNEXT routine to set the initial values of our outputs prior to the DO WHILE loop.

❹ Since we are trying to build a single record to contain all four data elements, we have to condition the OUTPUT statement on the last one of these elements that occurs in the XML—which happens to be respondents.

❺ The RETAIN statement must be used in order to keep all of the variable values between each occurrence of the OUTPUT statement. Otherwise, the DO loop will dump their values between each iteration.

Output 2.17: Sample of Extracted Data

Obs	ReleaseNumber	ReleaseDate	Respondents	URL
1	34-61262	Dec. 30, 2009	Stephen C. Gingrich	http://www.sec.gov/litigation/admin/2009/34-61262.pdf
2	34-61256	Dec. 30, 2009	Gabelli Funds LLC	http://www.sec.gov/litigation/admin/2009/34-61256.pdf
3	34-61255	Dec. 30, 2009	Gabelli Funds LLC	http://www.sec.gov/litigation/admin/2009/34-61255.pdf
4	34-61252	Dec. 29, 2009	Banc One Investment Advisors Corporation and Mark A. Beeson	http://www.sec.gov/litigation/admin/2009/34-61252.pdf
5	34-61247	Dec. 29, 2009	Jeffrey C. Young	http://www.sec.gov/litigation/admin/2009/34-61247.pdf
6	33-9101	Dec. 28, 2009	Bahram A. Jafari and Mountain Resources, Inc.	http://www.sec.gov/litigation/admin/2009/33-9101.pdf
7	34-61245	Dec. 28, 2009	Anna M. Baird, CPA	http://www.sec.gov/litigation/admin/2009/34-61245.pdf
8	34-61243	Dec. 28, 2009	Customer Sports, Inc., General Magic, Inc., Leonidas Films. Inc. (n/k/a Consolidated Pictures Group, Inc.), SportsPrize Entertainment, Inc., U.S. Interactive. Inc., and USA Biomass Corp.	http://www.sec.gov/litigation/admin/2009/34-61243.pdf
9	33-9100	Dec. 22, 2009	Applied Minerals, Inc. (Formerly Known as Atlas Mining Company)	http://www.sec.gov/litigation/admin/2009/33-9100.pdf
10	34-61210	Dec. 18, 2009	Inviva, Inc. and Jefferson National Life Insurance Company	http://www.sec.gov/litigation/admin/2009/34-61210.htm
11	34-61209	Dec. 18, 2009	CIHC, Inc., Conseco Services, LLC, and Conseco Equity Sales, Inc.	http://www.sec.gov/litigation/admin/2009/34-61209.htm
12	34-61208	Dec. 18, 2009	Cornerstone Capital Management, Inc., and Laura Jean Kent	http://www.sec.gov/litigation/admin/2009/34-61208.htm
13	33-9097	Dec. 18, 2009	ICAP Securities USA LLC, Ronald A. Purpora, Gregory F. Murphy, Peter M. Agola, Ronald Boccio, Kevin Cunningham, Donald E. Hoffman, Jr., and Anthony Parisi	http://www.sec.gov/litigation/admin/2009/33-9097.pdf
14	34-61199A	Dec. 17, 2009	Bear Wagner Specialists LLC, Fleet Specialist, Inc., LaBranche & Co. LLC, Spear, Leeds & Kellogg Specialists LLC, Van der Moolen Specialists USA, LLC, Performance Specialist Group LLC, and SIG Specialists, Inc.	http://www.sec.gov/litigation/admin/2009/34-61199a.pdf
15	33-9096	Dec. 17, 2009	Ernst & Young LLP	http://www.sec.gov/litigation/admin/2009/33-9096.pdf

As we can see in Output 2.17, the code effectively captured the four elements out of our XML source. This approach is generalizable to many other hierarchical file types such as HTML and should be interesting to explore.

Homework

1. Rearrange the ReleaseDate field to look like a different standard SAS date format.
2. Create a variable named Count that provides the number of Respondents on each row. This will be both fun and tricky.
3. Enhance the current RegEx pattern to force a match of a closing tag with the same name. There are two occurrences in the existing output where a formatting tag called SUB is embedded in the Respondents field. Our existing code stops on the closing tag for SUB instead of on the closing tag for `respondents`. The fix for this is not difficult, but requires more code than you might anticipate.

2.4.3 Search and Replacement

The specific needs for search and replace functionality can vary greatly, but nowhere is this capability more necessary than for PII redaction. Redacting PII is a frequent concern in the public sector, where information sharing between government agencies or periodic public information release is often mandated. We revisit the data from our cleansing and standardization example here since it includes the kinds of information that we would likely want to redact. However, in an effort to make this more realistic, the example data set that we used in Section 2.4.2 has been exported to a text file. We want to know how to perform this task on any data source, from the highly structured to the completely unstructured. We have already worked with structured data sources for this technique, so exploring unstructured data sources is a natural next step.

Now, before we get into how to perform the redaction, it is worth showing how the TXT file was created. Despite knowing how it is created in advance, we want to behave as though we have no knowledge of its construction in order to ensure that we are creating reasonably robust code.

You can see in the code snippet below that we simply take the resulting data set from Section 2.4.2, FinalPII, as input via the SET statement. Next, we use the FILE statement to create the TXT file reference. Once the FILE statement is used, the following PUT statement automatically writes the identified variables to it.

```
data _NULL_;
set finalpii;
file 'F:\Unstructured Data Analysis\Chapter_2_Example_Source\
FinallPII_Output.txt';
put surname firstname ssn dob phonenumber street city state zip;
run;
```

Output 2.18 shows the output provided by this code. As we can see, the structure is largely removed, though not entirely gone. This allows us to more closely approximate what you might encounter in the real world (which could be PII stored in a Microsoft Word file).

Output 2.18: PII Raw Text

```
SMITH JAMES 539-71-9216 12/17/1986 2-746-475-4589 1776 D Street NW Washington DC 20006
JOHNSON JOHN 189-03-1020 3/25/1981 0-439-270-9250 1600 Pennsylvania Avenue NW Washington DC 20500
WILLIAMS WILLIAM 971-45-0631 10/2/1986 6-281-794-3626 600 14th Street NW Washington DC 20005
JONES DAVID 277-05-1098 8/9/1985 9-349-208-5935 1321 Pennsylvania Avenue NW Washington DC 20004
BROWN THOMAS 123-22-9494 1/26/1980 3-287-870-2874 2470 Rayburn Hob Washington DC 20515
DAVIS CHRISTOPHER 688-49-1392 5/20/1992 6-639-100-7721 101 Independence Avenue SE Washington DC 20540
MILLER DANIEL 675-36-2461 9/9/1977 8-277-323-9564 511 10th Street NW Washington DC 20004
WILSON PAUL 304-59-8869 5/18/1980 6-863-034-6857 450 7th Street NW Washington DC 20004
MOORE MARK 102-01-9574 7/29/1991 5-697-801-1886 2 15th Street NW Washington DC 20007
TAYLOR DONALD 439-72-6850 2/1/1982 3-019-416-1550 3700 O Street NW Washington DC 20057
ANDERSON GEORGE 116-12-3837 8/10/1983 0-036-200-7891 3001 Connecticut Avenue NW Washington DC 20008
THOMAS KENNETH 549-54-0433 5/22/1976 1-705-466-8443 3101 Wisconsin Avenue NW Washington DC 20016
JACKSON EDWARD 461-17-1160 9/6/1977 7-841-664-8908 800 Florida Avenue NE Washington DC 20002
WHITE ANTHONY 374-43-5208 5/7/1979 8-451-939-9401 1 First Street NE Washington DC 20543
HARRIS KEVIN 877-45-2254 6/10/1984 7-690-620-4418 600 Independence Avenue SW Washington DC 20560
MARTIN MATTHEW 436-07-9380 10/29/1975 2-623-941-6074 10th Street and Constitution Avenue NW Washington DC 20560
THOMPSON LINDA 500-98-4809 3/10/1983 6-564-897-1662 555 Pennsylvania Avenue NW Washington DC 20001
GARCIA MARGARET 772-03-0744 10/16/1992 5-894-411-7166 1000 5th Avenue New York NY 10028
MARTINEZ LISA 426-25-2712 8/18/1978 1-167-511-7529 64th Street and 5th Avenue New York NY 10021
ROBINSON KAREN 761-09-3862 10/18/1974 5-395-540-6931 10 Lincoln Center Plaza New York NY 10023
CLARK DONNA 131-47-7120 11/3/1991 7-781-276-1544 Pier 86 W 46th Street and 12th Avenue New York NY 10036
RODRIGUEZ CAROL 280-95-9464 12/28/1985 9-737-785-2677 350 5th Avenue New York NY 10118
LEWIS SHARON 896-33-5968 1/31/1975 0-918-040-2361 405 Lexington Avenue New York NY 10174
LEE MICHELLE 735-62-9285 10/3/1984 9-488-783-8608 1 Albany Street New York NY 10006
WALKER KIMBERLY 109-02-1649 9/17/1990 7-823-096-2389 1585 Broadway New York NY 10036
```

Now that we have an unstructured data source to work with, we can create the code to redact all sensitive data elements. The challenge of doing this effectively is that we can't depend on the structure of surrounding text to inform the redaction decisions of our code. For this reason, it is important that our code takes great care to ensure that we properly detect the individual elements before redacting them.

Below is the code that performs our redaction of the PII elements SSN, DOB, PhoneNumber, Street, City, and Zip. Now, many organizations are allowed to publish small amounts of information that individuals authorize in advance, such as city or phone number. However, we're focusing on how to redact all of them, because keeping a select few is easy. In addition to performing the redaction steps, we also output the redacted text to a new TXT file. Again, this is an effort to support a realistic use of these techniques. For instance, many organizations keep donor or member information in text files that are updated by administrative staff. There are valid reasons for sharing portions of that information either internally or with select external entities, but doing so must be undertaken with great care. Thus, it is useful for such files to be automatically scrubbed prior to being given a final review and then shared with others.

Note: While the following code is more realistic than what we developed earlier in the chapter, it still needs to be improved for robust, real-world applications. The homework for this chapter has some suggestions, but there is always room for additional refinement.

```
data _NULL_;
infile 'F:\Unstructured Data Analysis\Chapter_2_Example_Source\F
inallPII_Output.txt' length=linelen lrecl=500 pad;
varlen=linelen-0;
input source_text $varying500. varlen;
Redact_SSN = PRXPARSE('s/\d{3}\s*-\s*\d{2}\s*-\s*\d{4}/REDACTED/o');   ❶
Redact_Phone = PRXPARSE('s/(\d\s*-)?\s*\d{3}\s*-\s*\d{3}\s*-
\s*\d{4}/REDACTED/o');
Redact_DOB = PRXPARSE('s/\d{1,2}\s*\/\s*\d{1,2}\s*\/\s*\d{4}/REDACTED/o');
Redact_Addr =
PRXPARSE('s/\s+(\w+(\s\w+)*\s\w+)\s+(\w+\s*\w+)\s+(\w+)\s+((\d{5}\s*?-
\s*?\d{4})|\d{5})/ REDACTED, $4 REDACTED/o');
CALL PRXCHANGE(Redact_Addr,-1,source_text);   ❷
CALL PRXCHANGE(Redact_SSN,-1,source_text);
CALL PRXCHANGE(Redact_Phone,-1,source_text);
CALL PRXCHANGE(Redact_DOB,-1,source_text);

file 'F:\Unstructured Data Analysis\Chapter_2_Example_Source\
RedactedPII_Output.txt';
put source_text;                               ❸
run;
```

❶ We create four different RegEx_ID's associated with the different PII elements that we want to redact from our source file—SSN, PhoneNumber, DOB, Street, City, and Zip.

❷ We use the CALL PRXCHANGE routine to apply the four different redaction patterns in sequence.

❸ Using the FILE statement, we create an output TXT file for writing our resulting text changes to. Since we overwrote the original text using the CALL PRXCHANGE routine (i.e., changes were inserted back into source_text), we need to output only the original variable, source_text, with the PUT statement.

Figure 2.2: Redacted PII Data

```
SMITH JAMES REDACTED REDACTED REDACTED REDACTED, DC REDACTED
JOHNSON JOHN REDACTED REDACTED REDACTED REDACTED, DC REDACTED
WILLIAMS WILLIAM REDACTED REDACTED REDACTED REDACTED, DC REDACTED
JONES DAVID REDACTED REDACTED REDACTED REDACTED, DC REDACTED
BROWN THOMAS REDACTED REDACTED REDACTED REDACTED, DC REDACTED
DAVIS CHRISTOPHER REDACTED REDACTED REDACTED REDACTED, DC REDACTED
MILLER DANIEL REDACTED REDACTED REDACTED REDACTED, DC REDACTED
WILSON PAUL REDACTED REDACTED REDACTED REDACTED, DC REDACTED
MOORE MARK REDACTED REDACTED REDACTED REDACTED, DC REDACTED
TAYLOR DONALD REDACTED REDACTED REDACTED REDACTED, DC REDACTED
ANDERSON GEORGE REDACTED REDACTED REDACTED REDACTED, DC REDACTED
THOMAS KENNETH REDACTED REDACTED REDACTED REDACTED, DC REDACTED
JACKSON EDWARD REDACTED REDACTED REDACTED REDACTED, DC REDACTED
WHITE ANTHONY REDACTED REDACTED REDACTED REDACTED, DC REDACTED
HARRIS KEVIN REDACTED REDACTED REDACTED REDACTED, DC REDACTED
MARTIN MATTHEW REDACTED REDACTED REDACTED REDACTED, DC REDACTED
THOMPSON LINDA REDACTED REDACTED REDACTED REDACTED, DC REDACTED
GARCIA MARGARET REDACTED REDACTED REDACTED REDACTED, NY REDACTED
MARTINEZ LISA REDACTED REDACTED REDACTED REDACTED, NY REDACTED
ROBINSON KAREN REDACTED REDACTED REDACTED REDACTED, NY REDACTED
CLARK DONNA REDACTED REDACTED REDACTED REDACTED, NY REDACTED
RODRIGUEZ CAROL REDACTED REDACTED REDACTED REDACTED, NY REDACTED
LEWIS SHARON REDACTED REDACTED REDACTED REDACTED, NY REDACTED
LEE MICHELLE REDACTED REDACTED REDACTED REDACTED, NY REDACTED
WALKER KIMBERLY REDACTED REDACTED REDACTED REDACTED, NY REDACTED
```

As we can see in the resulting file of redacted output (Figure 2.2), only the individual's name and state are left. The redacting clearly worked, but in this context the resulting information might not be the most readable. How can we achieve the same goal while making the output easier to read? It's simple. Instead of inserting REDACTED, we can insert "" (i.e., nothing), which effectively deletes the text. Try it out and see what happens.

Homework

1. Update the RegEx patterns to allow City to be shown (tricky with two-word names like New York).
2. Incorporate the results of Section 2.4.1, Homework item 5 so that the Census tract can be displayed.
3. Use the random PII generator in Appendix A to incorporate an entirely new field to then display this output in.
4. Make this code more robust by incorporating zero-width metacharacter concepts such as word boundaries (\b) to ensure that word edges are identified properly.

2.5 Summary

In this chapter, we have explored the PRX suite of functions and call routines available in SAS for implementing RegEx patterns, and a few examples of practical uses for them. They collectively provide tremendous capability, enabling advanced applications.

As we have seen throughout the chapter, PRX functions and call routines *cannot* replace well-written RegEx patterns, despite providing incredible functionality. Attempting to leverage functions and call routines with poorly written RegEx patterns is like trying to drive a sports car with no fuel.

Also, while the PRX functions and call routines represent flexible, powerful capabilities to be leveraged for a wide variety of applications—basic and advanced—they often cannot stand alone. It is important to leverage them in conjunction with other elements of SAS to develop robust code; a fact that we have merely had a glimpse of in this chapter. For instance, some very advanced RegEx applications benefit from the use of MACRO programming techniques (beyond the scope of this book).

I hope this helps you to get more comfortable with RegEx, and primed to apply these concepts to entity extraction in Chapter 4. You have all the basic tools in place to make truly useful, robust SAS programs that leverage regular expressions. However, as promised from the outset, we are going to really pull everything together in the coming chapters to extract and manipulate entity references.

----

[1] McCallister, Erika, Tim Grance, and Karen Scarfone, "Special Publication 800-122: Guide to Protecting the Confidentiality of Personally Identifiable Information (PII)," *National Institute of Standards and Technology,* April 2010, http://csrc.nist.gov/publications/nistpubs/800-122/sp800-122.pdf.

[2] SAS Institute Inc., "BASE SAS DATA Step: Perl Regular Expression Debug Support," *SAS Support,* https://support.sas.com/rnd/base/datastep/perl_regexp/regexp.debug.html, (accessed August 29, 2018).

[3] The source file was downloaded manually from the SEC website in XML format.
SEC, "SEC Administrative Proceedings for 2009," *U.S. Securities and Exchange Commission,* http://www.sec.gov/open/datasets/administrative_proceedings_2009.xml (accessed August 29, 2018).

Chapter 3: Entity Resolution Analytics

3.1 Introduction

Organizations in every domain need to properly manage information about individuals and assets being tracked within their data infrastructure. These individuals and assets are generically referred to by analytics professionals as "entities," because the term can apply to people, places, or things.

Entity is defined as something that has a real existence.

An entity can truly be any real person, place, or thing that we want to track. Below are just a few examples.

- Customer
- Mobile phone
- Contract
- House listed by a realtor
- Fleet vehicle
- Bank Account
- Employee

The specific business need will determine what entities, and associated data elements, are being tracked by an organization's data infrastructure. The process of tracking those entities will be customized accordingly. Regardless of an entity type, all data and information about an entity within databases, free text, or other discoverable sources are known as "entity references."

Entity reference: information or data about the real-world person, place, or thing being referred to.

For example, an entity reference could include a customer profile that appears as a single record in a customer relationship management database, an online real-estate listing for a specific house, or a phone number listed in a billing statement.

As we can see in the example below, an entity—in this case, SAS Institute Inc.—can have incomplete reference information, making the task of connecting that reference to the entity quite challenging. This is an important element of entity resolution, and one we will spend significant time on in the upcoming chapters.

Figure 3.1: Entity Reference Example[1]

> Using SAS tools for data analysis, the analytics department takes its skills to different departments within the company.

Organizations of all kinds struggle with tracking and analyzing entities in different ways, but the underlying technology and analytical tools to overcome their challenges remain the same. Thus, the discussion throughout this book will provide examples across several domains, while keeping the methodology very general.

However, as we will see, there are important decisions to be made throughout this methodology that will drive effective implementation of entity resolution for the particular business context. So, thoroughly understanding the business context is incredibly important, and will ultimately shape the success or failure of entity resolution in practice.

3.2 Defining Entity Resolution

When sifting through the various entity references within institutional data sources, we have to determine which entities are being referred to, and do so in an effective manner. This requires that we apply a rigorous, repeatable process to evaluate each entity reference against every other reference (i.e., pairwise comparison) in our data sources. This rigorous process is called "entity resolution."

> **Entity resolution** is the act of determining whether or not two entity references in a data system are referencing the same entity.

The concept is actually simple—we compare uniquely identifying attributes of both entity references (e.g., Social Security numbers (SSN)) to determine a match, and repeat this kind of comparison for every entity reference identified in our data sources. However, the implementation is much more complex due to changing business needs, data quality issues, and ever-changing entities—making matches less straightforward. In other words, real-world implementation of entity resolution is difficult, and the particular business application dramatically impacts the kinds of challenges we will need to overcome.

A robust method for entity resolution is critical for large-scale business operations improvement, and enables a wide variety of applications—some examples of which are below:

Domain	Application
Retail	Social Media Ad Campaign Analysis and Planning
Finance	Know Your Customer (KYC), Insider Trading
Government	Insider Threat Detection
Insurance	Fraud, Waste, and Abuse Prevention

I will revisit some of these and other examples over the coming chapters to demonstrate the variety of ways high-quality entity resolution can impact an organization's day-to-day operations.

3.3 Methodology Overview

As we will see during the remainder of this book, the steps involved in applying Entity Resolution (ER) have evolved in recent years to imply far more than just the simple act of resolving two entities. Real-world applications of ER have precipitated the need for an analytical framework that contemplates the end-to-end steps for acquiring data, cleansing it, resolving the entity references, analyzing it for the specific need, and managing the resulting reference linkages. By establishing a robust framework, we are able to apply ER to numerous subject domains, capable of solving a variety of problems. I and some other authors in the industry refer to this framework as Entity Resolution Analytics (ERA).[2]

The phases that we will discuss for ERA have been sufficiently generalized to support the application of this methodology to a wide variety of data and domain types. Applying the high-level phases described below will have some underlying variation (and nuance not depicted here), depending on the specific problem being solved; however, the major steps shown in Figure 3.2 will not change.

Figure 3.2: Overview of the ERA Flow

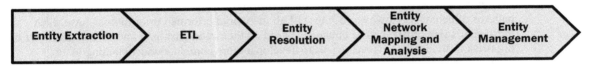

3.3.1 Entity Extraction

We begin with entity extraction in both the business and technical contexts as the initial set of tasks for any ERA project. In the business sense, we must define what kinds of entities we want to extract and for what purpose. And the technical elements of the project have to be developed, or put in place to support the identified needs.

> **Example:** A hedge fund manager is researching companies.
>
> The manager wants to better understand companies currently in her portfolio, as well as a few she is evaluating for inclusion in the fund. She likely has at her disposal software for financial professionals that gives her access to incredible volumes of information about the companies of interest, including news articles. This is possible only because that software has sophisticated algorithms for *extracting* and *resolving* the entity references in those articles effectively. In so doing, the fund manager is able to gain access to information that enables her to make well-informed decisions.

3.3.2 Extract, Transform, and Load

Extract, Transform, and Load (ETL) are the classic processes for changing and moving data in Relational Database Management Systems (RDBMS). ETL processes enable us to take data from its raw form and migrate it through database systems in a repeatable way to ensure a dependable, predictable approach to moving and shaping data for our end use. In addition to pulling structured sources for an ERA project, ETL will be performed on the staging tables of entity references generated by the entity extraction phase.

As with every phase of the ERA framework, this is technology function driven by business needs and best practices. Robust business processes will ensure the proper application of ETL technology functions to ensure the level of quality and consistency that we need for each and every project being executed.

3.3.3 Entity Resolution

The next step in our process is entity resolution. This step is immediately following ETL since all the data should now be prepared and staged in the appropriate form to actually begin the process of evaluating each entity against each other entity identified during the extraction phase.

3.3.4 Entity Network Mapping and Analysis

After ER has been completed, we have to then make decisions about what resulting linkages are kept and which are thrown out, as well as what we can do with the results. I'm calling this process "entity network mapping and analysis" since the links that we have we are effectively *mapping* out the network of entities we will want to *analyze*. As will be discussed in the upcoming chapters, we have a lot of flexibility in determining thresholds for network linkages to be retained. This process is driven by business decisions made at the outset of the project. The variety of analysis that could be performed here is quite broad; however, I will execute some common analytical approaches that are achievable in the scope of this book.

> **Example:** A bank is attempting to identify fraud.
>
> Each individual in the bank's data stores will be analyzed for risk of fraud based on historical patterns of behavior. However, because of all the work performed prior to this phase, each individual's risk profile can be enhanced based on their relationship to known bad actors, or other high-risk individuals. This enables us to develop a more complete understanding of each individual's fraud risk.

3.3.5 Entity Management

For long-term monitoring of the entities tracked in your data stores, it is important to have a gold standard or baseline understanding of said entities. So, after an initial set of verified entity references has been developed, it then becomes the baseline against which we compare new entity references. As genuinely new entities are identified through these new references, we add them to our gold standard database. Redundant references are ignored, while references that augment our understanding of existing entities will be used to edit the database.

3.4 Business Level Decisions

Every company or agency has a different set of procedures, and those must be followed. But the key decisions that need to be made for an ERA project fit neatly into any existing management framework. So, let's go through key elements that you need to nail down before jumping into the technical elements of *how* ERA is executed.

3.4.1 Establish Clear Goals

Determine the ultimate goal for an ERA project. *What business goal(s) are you trying to achieve?* If you are a manager of such a project, you must be able to identify the goal(s). If you are the person implementing the technical aspects of an ERA project, you still must elicit this information from your management chain. Documentation of the goal(s) ensures there is a record of common understanding. Without a commonly agreed upon goal(s), projects easily lose focus and success criteria are rarely achieved. Perceived "failure" of these efforts generally stems from a fundamental mismatch in understanding of what is feasible in the period of time allotted for the project.

3.4.2 Verify Proper Data Inventory

Identify all the data sources available for the project, and determine whether the high-level goals are even realistic prior to starting into any project. Once the feasibility is well-known, the specific elements from which entities will be extracted, and the quality of those sources, will need to be determined. Entity references being pulled from structured data will also be documented, but execution of that plan will occur during the ETL phase of the ERA framework. This is a critical item that can't be emphasized enough, and goes back to goals and expectation management. In many cases, you have structured elements that you are trying to enrich with unstructured sources. However, if you don't have some key pieces of structure data to leverage for combining with the unstructured references, the goals that you have established may never be realized.

3.4.3 Create SMART Objectives

SMART objectives are aligned to the overall goal of this project—and they are Specific, Measurable, Achievable, Relevant, and Time-bound. This is a direct result of the reasons that were already stated for project failure. We need to ensure that project goals are rationalized via project objectives that can be achieved in a time and performance window acceptable to the project sponsor. This is a critical step for any technical project like ERA as it is the nexus for project management and technical staff to thoroughly set expectations, and examine trade-offs jointly. Here are some notional examples of the things being defined for each of these elements.

> **S**pecific: Capture company names in news source X.
>
> **M**easurable: Use sample data set x to train capability and holdout set y for testing.
>
> **A**chievable: Attempt accuracy equal to known/advertised baseline (e.g., 75%).
>
> **R**elevant: The data source is the business section of a financial newspaper.
>
> **T**ime-bound: Complete project development within 90 days.

The particular business problem at hand may make some of these decisions quite obvious, but it is important to be very intentional, documenting every decision as you go through a project. You will want to refer back to that documentation weeks, months, or perhaps years later to understand why something is set up the way it is—or what decisions led to a particular development path for a solution. Hindsight is always perfect, making it is easy to ignore for factors that may seem less important years later while they were critical during project execution.

Note: An added benefit of this documentation is the opportunity to learn from past project failures and successes.

There is much more that could be said regarding project management best practices, but I don't want to go too deeply into that topic in this book. The above list is certainly not complete, but contains reminders that can't be emphasized enough, regardless of the management methodology that you are implementing (e.g., Agile).

3.4 Summary

Whether you are a practitioner of analytics, or a manager of analytics teams, I hope you find this information helpful in getting started down the road of effectively executing ERA projects with SAS. As I said before, we will explore the foundational aspects of ERA, with pointers about how you can expand your knowledge and capabilities to execute much larger scale projects of this type—we have to start somewhere. Now, let's have some fun!

---

[1] Groenfeldt, Tom, "Toyota Finance Uses Advanced Analytics to Improve Sales and Profits," *Forbes*, April 13, 2007, https://www.forbes.com/sites/tomgroenfeldt/2017/04/13/toyota-finance-uses-advanced-analytics-to-improve-sales-and-profits/#78d9bc655cb7 (accessed August 29, 2018).

[2] Talburt, John R., *Entity Resolution and Information Quality* (Burlington, MA: Morgan Kaufmann, 2010).

Chapter 4: Entity Extraction

4.1 Introduction

Figure 4.1: ERA Flow with Entity Extraction Focus

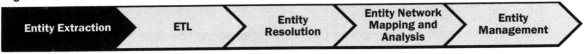

In order to identify the free text entity references we want to utilize for all subsequent resolution and analysis activities, we have to first determine the types of entities we want to understand more about, and in what data sources their references reside. This initial step in the ERA framework is known as "entity extraction."

    Entity extraction: the algorithmic identification of entity references in source data.

From a purely technical perspective, this phase of work is primarily applicable to unstructured and semi-structured data sources. However, from a business perspective, this initial phase of ERA is applicable to all data sources—a larger scope than what is implied by our definition.

As we shall discuss further, the underlying technical steps are quite different depending on the type of data (structured vs. unstructured), but the business-level decisions made at the beginning of this phase will inform each branch of work to follow. The decisions made will flow into the entity extraction efforts being undertaken during this phase for the unstructured and semi-structured data, as well as the ETL work taking place in the next phase for any structured entity reference sources.

The extraction of entity references from unstructured text is accomplished through something called text parsing, where text is algorithmically analyzed using previously determined rules. The process of developing these rules can be quite difficult and time-consuming, depending on the domain and approach. Due to the complexity of some rules, this effort may become a disproportionately large percentage of the total ERA effort—which is why private companies like SAS have developed sophisticated software to make this process easier, more accurate, and faster. It is also why context is incredibly important. As the context for an entity reference becomes more broad, the possibilities for what is "normal" in that source text becomes increasingly more difficult to assess or anticipate. As the source data context becomes more broadly defined, more sophisticated rules must be created in order to account for a greater number of possibilities. For example, medical reports will have a more narrow set of possible configurations of data elements and meanings, compared to a news article.

Regardless of the sources used for collecting data, the extracted entities will need to be placed into temporary data tables (staging tables) in preparation for the next step in our process.

Entity Extraction

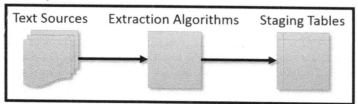

"But, what if my project doesn't have unstructured data elements?" you ask. Well, I would be both surprised and happy for you. And if you are as lucky as to deal only with structured sources, I would say you should still go through the business level decisions portion of this chapter to properly document the business goals and assumptions for the work being performed. Those decisions will need to be referenced later to help both technical and managerial staff stay on course as the project progresses.

Note: Context is incredibly important, and we are not going to "boil the ocean" for any project. Only the data sources necessary to achieving our goals will be selected for text parsing.

And before we go any further, I think it is important to note that we are not going to build something of science fiction movie fame. It is not realistic to dream of building some kind of omniscient machine that will perfectly and neatly extract every single kind of entity reference everywhere, and clean it for feeding our version of HAL 9000. Every ERA project will have a business or mission-specific goal, and to go beyond the scope of said goal will be counterproductive.

4.2 Business Context

Before we jump into the technical details of extracting entity references, I want to review some project execution concepts. Assuming that the data inventory has been completed (see Section 3.4.2), we already have an understanding of what is feasible with the available data sources. So, this provides the high-level context that we need to make the necessary decisions about the technical approach.

When extracting entity references from text, we have to be careful to track the origination of each reference for later use. Depending on the data source structure and context, as well as the nature of our subsequent analyses, we may need to retain very detailed indexing of each reference. For example, we may have word documents with dense information broken into paragraphs, and we want to later understand if two entities are related based purely on their "closeness" within the source text. There are many ways to identify entity relationships: this is merely one simple approach. Knowing that they are not only in the same document, but also in the same paragraph or sentence can greatly improve our assessment that they are related in some way. So, we would need to account for that in our indexing method when extracting the entity references.

There are more advanced techniques that get into natural language processing, but this is one basic way for identifying relationships without characterizing them.

As you can imagine, we have to think about those possible future analyses now, and establish the necessary indexing scheme for entity reference extraction upfront. Otherwise, we will have a lot of refactoring to perform much later in the project, which wastes precious time and money. So, just remember the sound management mindset of ***plan now or pay later***.

In order to get us going on the right path, I'm going to focus on developing the building blocks of entity reference extraction that will enable you to put together relatively sophisticated extraction rules.

4.3 Scraping Text Data

There are many different potential sources of data for an ERA project, and I can't hope to cover them all (or even most of them) here. So, I want to give you the programmatic tools to grab realistic text data in a practical way. And it is to that end that I will show you a couple ways to scrape text data sources for entity extraction below.

But wait, this sounds like ETL! Well, it is technically in the realm of the data "Extraction" portion of ETL, but keep in mind that we are attempting to organize around a methodical business process for executing this kind of work—Entity Resolution Analytics or ERA. So, in order to *do* the entity extraction that we are covering in this chapter, we have to grab the raw source text first. I will get into the details of what is done with the results during our next chapter, when I walk you through the ETL processes, and ways to prepare the extracted entities for combination with structured data sources and analyses.

4.3.1 Webpage

It is often the case that we want to leverage publicly available, unstructured, or text data sources for analysis. So, I'm going to show you one way to easily grab data off of a website using PROC HTTP.

PROC HTTP is just one of several methods for acquiring data outside of the local file system or database. And it is an easy method for integration into Base SAS code, like we have used thus far. I will not go through all the different powerful options available in PROC HTTP, but just what we need for the subsequent examples. Please refer to the documentation for more details on all the additional ways to use PROC HTTP.[1]

In the below code, I start with a null DATA step to create a macro variable named FileRef that contains a dynamic file reference. Whenever we are extracting from text sources, it is a best practice to ensure the date of extraction is maintained. I have chosen to do this by embedding the extraction date in the file name that I created for the resulting file. Next, the FILENAME statement uses this macro variable to establish the file reference, Source. And finally, I use Source in PROC HTTP as the output file, using OUT=. Note the URL= statement identifies a link to the BBC News feed (The BBC allows all forms of reuse of their content so long as they are properly referenced.).

```
/*NULL DATA STEP to generate the macro variable with a dynamic file name.*/
data _NULL_;
call
symput('FileRef',"'C:\Users\mawind\Documents\SASBook\Examples\CorpNews"||put(Date
(),date9.)||".txt'");
run;
```

```
filename source &fileref;

/*Execution of PROC HTTP to extract from a web URL, and write to the file
designated above.*/
proc http
    url="http://feeds.bbci.co.uk/news/technology/rss.xml"
    out=source;
run;
```

After the above code runs, you will get a raw text file containing all of the XML content from the source (Figure 4.2 below).

Figure 4.2: Sample of TXT from BBC News RSS Feed

Depending on how the host (BBC News, in this case) decides to maintain that information, your refresh and indexing of it will vary. Luckily, in our example, we can use the <pubdate> tag embedded in the XML to help us track when each headline was published.

Note: You are not constrained to XML when using PROC HTTP. It is merely how a source that I'm allowed to reprint stores data. So, feel free to experiment with pages that don't use XML.

But much of the data contained within webpage XML files is unstructured text (some call this "semi-structured" data). Therefore, you want to be prepared for how to approach entirely unstructured text data—a much more voluminous portion of potential sources. Since we went through examples of processing tag languages in Chapter 1, I'm not going to repeat that here. I will instead focus on patterns for capturing entities in the free text portion of our raw source.

4.3.2 File System

Now, we often have significant unstructured or textual data sources available on a file system, which we also want to leverage for entity extraction, resolution, and analysis. In order to help you do this without any additional software (i.e., just Base SAS), I have written the below code using only Base SAS functions and features, in addition to examining only TXT files.

Note: This code neatly picks up where the PROC HTTP code left off. So, it is easy to imagine how that can be incorporated for a larger macro to grab large amounts of data from web sources.

I begin in the below code by creating a macro variable identifying the directory I want to search. Next, the DATA step creates a data set named Files that contains the variables Filename, File_no_ext, and Full_path as the final output. Within the DATA step, I use a series of directory functions in SAS to determine the number of files in the directory and their names. While most of the functions and statements in the below code are likely familiar, I have used callouts to highlight sections of some less commonly used elements.

```
%let dir=C:\Users\mawind\Documents\SASBook\Examples\;

data files (keep=filename file_no_ext full_path label="All Text Sources");
    length filename file_no_ext full_path $1000;
    rc=filename("dir","&dir");
    did=dopen("dir");              ❶
    if did ne 0 then do;
        do i=1 to dnum(did);       ❷
            filename=dread(did,i);
            file_no_ext=scan(filename,1,".");   ❸
            full_path=Cat(&dir,filename);
        if lowcase(scan(filename, -1, "."))="txt" then output;
        end;
    end;
    else do;
        put "ERROR: Failed to open the directory &dir";
        stop;
    end;
run;

proc sort data=files;
    by full_path;
run;

proc print data=files;
run;
```

❶ The DOPEN function takes the directory as an input. It then opens that directory, and creates an ID number greater than zero to identify it later.

❷ DNUM is used to determine the number of members there are in the supplied directory ID number. In this case, our directory ID is "did."

❸ Inside the =DO loop, I use DREAD to extract the member filename into the variable Filename.

After the code runs, you should now have a data set named Files with the three variables, Filename, File_no_ext, and Full_path. The output from my example folder is shown in Output 4.1.

Output 4.1: PROC PRINT Output of WORK.FILES

Obs	filename	file_no_ext	full_path
1	CorpNews01OCT2017.txt	CorpNews01OCT2017	C:\Users\mawind\Documents\ERA_SASBook\Examples\CorpNews01OCT2017.txt
2	PII_text.txt	PII_text	C:\Users\mawind\Documents\ERA_SASBook\Examples\PII_text.txt
3	sec_data.txt	sec_data	C:\Users\mawind\Documents\ERA_SASBook\Examples\sec_data.txt

Now, we have some more advanced steps to perform in order to complete the process of extracting the text, but bear with me; it's worth the extra effort. This next chunk of code takes the file names seen in our data set output above, and builds macro variables with them. This sets up the flexibility that we need to process folders with any unknown number of files.

In the below code, I create macro variables for the file names and paths, which are delineated by spaces. Again, most of this code is straightforward, but see the highlights listed below.

```
data _NULL_;
set files end=last;
length myfilepaths $500. myfiles $50.;
if _N_=1 then do;
myfilepaths="'"||strip(full_path)||"'";      ❶
myfiles="'"||strip(file_no_ext)||"'";
end;
else do;
myfilepaths="'"||strip(full_path)||"'"||' '||strip(myfilepaths);
myfiles="'"||strip(file_no_ext)||"'"||' '||strip(myfiles);
end;
if last then do;
    call symputx('filelist',myfilepaths);    ❷
    call symputx('filenames',myfiles);
    call symputx('filecount',_N_);
end;
put myfilepaths myfiles;
retain myfilepaths myfiles;
run;
```

❶ The IF and ELSE statements combine to create string variables containing full path names and just file names. Notice the double quotation marks surround the single quotation marks on either side of each instance of a full path name. That is done to provide the INFILE statement fully delineated path names in our next chunk of code, which does the actual processing.

❷ The CALL SYMPUTX statements here create the final macro variables used in the next chunk of code for actually ingesting and parsing the data.

In order to provide a neat completion to this process, I want to include patterns for processing the data identified in the files. So, I will do that in the next section before bringing all the pieces together in Section 4.5. It is there that I will show you how we finally bring the web and local file data sources in our file system together with a custom text parsing macro.

4.4 Basic Entity Extraction Patterns

Now, I will begin putting together extraction patterns for specific types of entity references. This will create a foundation upon which you can build increasingly sophisticated extraction routines. Some of the elements discussed below will look quite familiar from Chapter 2, but it is worth reviewing them with this new aim in mind.

Caution: This is not meant to be a prescriptive listing. Your source data could override the precise patterns developed herein, but these should give you a starting point for future work. Remember, *context* is critical.

For Sections 4.4.1–4.4.3 below, I will be revisiting the PII data from Chapter 2 (see Figure 4.3 as a reminder), demonstrating methods for parsing and tracking those references. And I will use data scraped in Section 4.3 for our examples in Sections 4.4.4 and 4.4.5 below.

As a reminder, below is a file of our PII output from the random PII generator code (see Appendix A). Although this data is obviously artificial due to privacy concerns of publishing real PII data, I hope you can see the connection between these artificial examples and the real-world implementation of parsing such data.

Figure 4.3: Source File of Example PII Data

```
JAMES SMITH 5(109)673-7773 681-28-5055 12/17/1986 1776 D St NW, Washington, DC 20006
WILLIAM JOHNSON +4.814.554.3535 876-08-3421 3/25/1981 1600 Pennsylvania Ave NW, Washington, DC 20500
DAVID WILLIAMS 9-303-370-0447 341-42-1097 10/2/1986 600 14th St NW, Washington, DC 20005
CHARLES JONES 9-355-817-1288 798-17-0233 8/9/1985 1321 Pennsylvania Ave NW, Washington, DC 20004
JOSEPH BROWN 0-732-432-4989 604-98-8427 1/26/1980 2470 Rayburn Hob, Washington, DC 20515
CHRISTOPHER DAVIS +9.476.408.8932 395-83-0675 5/20/1992 101 Independence Ave SE, Washington, DC 20540
DANIEL MILLER 0(816)797-3833 423-05-0805 9/9/1977 511 10th St NW, Washington, DC 20004
PAUL WILSON 1-279-859-8916 327-26-7779 5/18/1980 450 7th St NW, Washington, DC 20004
GEORGE MOORE 9-735-840-8770 368-57-4711 7/29/1991 2 15th St NW, Washington, DC 20007
KENNETH TAYLOR 9(970)711-3846 732-62-4748 2/1/1982 3700 O St NW, Washington, DC 20057
STEVEN ANDERSON 4-262-255-0151 313-97-2129 8/10/1983 3001 Connecticut Ave NW, Washington, DC 20008
BRIAN THOMAS +2.290.379.0294 835-23-6538 5/22/1976 3101 Wisconsin Ave NW, Washington, DC 20016
RONALD JACKSON 7(584)093-1460 374-07-5121 9/6/1977 800 Florida Ave NE, Washington, DC 20002
ANTHONY WHITE 6(345)333-6042 561-03-0379 5/7/1979 1 First St NE, Washington, DC 20543
MARY HARRIS 5-226-333-2137 988-72-2288 6/10/1984 600 Independence Ave Sw, Washington, DC 20560
LINDA MARTIN +6.767.725.5936 905-98-2450 10/29/1975 10th St. & Constitution Ave. NW, Washington, DC 20560
ELIZABETH THOMPSON +4.861.503.2449 586-66-3374 3/10/1983 555 Pennsylvania Ave NW, Washington, DC 20001
JENNIFER GARCIA +1.439.239.5237 410-90-8822 10/16/1992 1000 5th Ave, New York, NY 10028
MARIA MARTINEZ +9.650.302.3781 411-86-6852 8/18/1978 64th St and 5th Ave, New York, NY 10021
NANCY ROBINSON 6(989)229-2108 059-25-4637 10/18/1974 10 Lincoln Center Plaza, New York, NY 10023
BETTY CLARK +5.547.927.9624 290-15-9219 11/3/1991 Pier 86 W 46th St and 12th Ave, New York, NY 10036
CAROL RODRIGUEZ 1(006)817-6827 155-35-3642 12/28/1985 350 5th Ave, New York, NY 10118
MICHELLE LEWIS 7-355-619-9245 257-70-8020 1/31/1975 405 Lexington Ave, New York, NY 10174
LAURA LEE 7-218-696-8153 555-21-3974 10/3/1984 1 Albany St, New York, NY 10006
SARAH WALKER +0.934.916.4094 571-85-3210 9/17/1990 1585 Broadway, New York, NY 10036
```

I ran the below code to parse the PII text for Sections 4.4.1–4.4.3. The code should be familiar at this point. I am merely taking the information matched by the pattern inside of the source PII text file, and printing that information to the log. As you can see below, I go through the results for each pattern separately in the sections below.

```
data _NULL_;
infile 'C:\Users\mawind\Documents\SASBook\Examples\PII_Text.txt' length=linelen
lrecl=500 pad;
varlen=linelen-0;
input source_text $varying500. varlen;

/*SSN Pattern*/
*Pattern_ID = PRXPARSE("/\b\d{3}\s*-\s*\d{2}\s*-\s*\d{4}\b/o");

/*Phone Pattern*/
Pattern_ID = PRXPARSE("/(\+?\d\s*(-|\.|\())?\s*?\d{3}\s*(-|\.|\)))\s*\d{3}\s*(-
|\.)\s*\d{4}/o");

/*Address Pattern*/
*Pattern_ID =
PRXPARSE("/\s+(\w+(\s\w+)*\s\w+),?\s+(\w+\s*\w+),?\s+(\w+),?\s+((\d{5}\s*-
\s*\d{4})|\d{5})/o");

CALL PRXSUBSTR(Pattern_ID, source_text, position, length);
if position ^= 0 then
   do;
      match=substr(source_text, position, length);
      put match:$QUOTE. "found in" source_text:$QUOTE.;
      put;
   end;

run;
```

4.4.1 Social Security Number

Social Security numbers are one of the easiest items to parse from free text as their standard appearance tends to be quite structured. They certainly could appear differently than described below, but that could be known only by taking sample data.

The pattern below requires that 3 digits are followed by a hyphen, then 2 digits and another hyphen, and 4 digits; each element can have white space between it and the next, but nothing else. And the entire pattern has to be bookended by anything that is not a digit character. This ensures we do not obtain errant matches in complex text.

SSN Pattern: "/\b\d{3}\s*-\s*\d{2}\s*-\s*\d{4}\b/o"

Output 4.2: Sample Log Output for SSN Pattern

```
"681-28-5055" found in
"JAMES SMITH 5(109)673-7773 681-28-5055 12/17/1986 1776 D St NW, Washington, DC 20006"

"876-08-3421" found in
"WILLIAM JOHNSON +4.814.554.3535 876-08-3421 3/25/1981 1600 Pennsylvania Ave NW, Washington, DC 20
500"

"341-42-1097" found in
"DAVID WILLIAMS 9-303-370-0447 341-42-1097 10/2/1986 600 14th St NW, Washington, DC 20005"

"798-17-0233" found in
"CHARLES JONES 9-355-817-1288 798-17-0233 8/9/1985 1321 Pennsylvania Ave NW, Washington, DC 20004"
```

4.4.2 Phone Number

Phone numbers are also relatively straightforward to extract from text data sources. The pattern discussed below accounts for a wide variety of US and international formats for telephone numbers.

This instantiation of the phone number capture pattern ensures that we capture 10-digit telephone numbers with optional international codes, and a variety of number delimiters (e.g., dashes are common in the US, while periods are common in Europe).

Phone Pattern: "/(\+?\d\s*(-|\.|\())?\s*?\d{3}\s*(-|\.|\)))\s*\d{3}\s*(-|\.)\s*\d{4}/o"

Output 4.3: Sample Log Output for Phone Number Pattern

```
"5(109)673-7773" found in
"JAMES SMITH 5(109)673-7773 681-28-5055 12/17/1986 1776 D St NW, Washington, DC 20006"

"+4.814.554.3535" found in
"WILLIAM JOHNSON +4.814.554.3535 876-08-3421 3/25/1981 1600 Pennsylvania Ave NW, Washington, DC 20
500"

"9-303-370-0447" found in
"DAVID WILLIAMS 9-303-370-0447 341-42-1097 10/2/1986 600 14th St NW, Washington, DC 20005"

"9-355-817-1288" found in
"CHARLES JONES 9-355-817-1288 798-17-0233 8/9/1985 1321 Pennsylvania Ave NW, Washington, DC 20004"
```

4.4.3 Address

Accounting for international formats here is much more difficult to do for addresses than for phone numbers (above). However, the below pattern addresses a variety of US addresses. You will need to adapt this to grab some address types such as those including PO boxes, or in very different sources. As always, test this on your source data. If we vary the source data just a bit, this pattern would break; so, sample your source data, and analyze the results carefully before implementing.

Also, each country's postal system has nuances that need to be accounted for with patterns specific to that country. So, it is important to remember this if you are parsing data from multiple countries.

Address Pattern: "/\s+(\w+(\s\w+)*\s\w+),?\s+(\w+\s*\w+),?\s+(\w+),?\s+((\d{5}\s*-\s*\d{4})|\d{5})/o"

Output 4.4: Sample Log Output for Address Pattern

```
" 1776 D St NW, Washington, DC 20006" found in
"JAMES SMITH 5(109)673-7773 681-28-5055 12/17/1986 1776 D St NW, Washington, DC 20006"

" 1600 Pennsylvania Ave NW, Washington, DC 20500" found in
"WILLIAM JOHNSON +4.814.554.3535 876-08-3421 3/25/1981 1600 Pennsylvania Ave NW, Washington, DC 20
500"

" 600 14th St NW, Washington, DC 20005" found in
"DAVID WILLIAMS 9-303-370-0447 341-42-1097 10/2/1986 600 14th St NW, Washington, DC 20005"

" 1321 Pennsylvania Ave NW, Washington, DC 20004" found in
"CHARLES JONES 9-355-817-1288 798-17-0233 8/9/1985 1321 Pennsylvania Ave NW, Washington, DC 20004"
```

The discussion in Sections 4.4.1–4.4.3 was a fun refresher on simple patterns, and how to apply them. Now, I will get back to the more challenging examples derived from the web data scraped back in Section 4.3.1. I have a chunk of code below that parses the RSS feed data for website names and corporation names.

Now, pay close attention to the corporation names pattern. This will be a tricky one to get clean, as we will explore in Section 4.4.5 below.

```
data RSSFeed_Data;
infile source length=linelen lrecl=500 pad;
varlen=linelen-0;
input source_text $varying500. varlen;

Corp_Pattern = "/(\b[A-Z]\w+\s[A-Z]\w+(\s[A-
Z]\w+)*\b)|(\w+\s+(\ucorp\b|\uinc\b|\uco\b))/o";
Website_Pattern = "/\b\w+\.(com|org|edu|gov)\b/o";

ExtractDate = put(Date(),date9.);

pattern_ID = PRXPARSE(Corp_Pattern);
start = 1;
stop = length(source_text);
CALL PRXNEXT(pattern_ID, start, stop, source_text, position, length);
   do while (position > 0);
      line=_N_;
      found = substr(source_text, position, length);
      put "Line:" _N_ found= position= length= ;
      output;
      CALL PRXNEXT(pattern_ID, start, stop, source_text, position, length);
      retain source_text start stop position length found;
   end;

keep ExtractDate line position length found;
run;
proc print data=Rssfeed_data;
run;
```

4.4.4 Website

Entire web addresses can get quite complicated, but we are generally interested in the root website rather than the entire address listed in our text source. There are clearly specific cases that run counter to this notion, such as mapping webpage hops by dissecting weblogs, but I will stick to the more general case here of website identification. Websites of many types are attributable to an entity of interest, such as a company or government agency, which is our aim in the general case. As you can see, we have many references to the BBC's website due to all the embedded links in the source page. This is something I will discuss cleaning up later.

Website Pattern: "/\b\w+\.(com|co|org|edu|gov|net)\b/o"

Output 4.5: Sample Website Pattern Results

Obs	ExtractDate	position	length	line	found
1	03OCT2017	23	8	3	purl.org
2	03OCT2017	72	8	3	purl.org
3	03OCT2017	130	6	3	w3.org
4	03OCT2017	189	9	3	yahoo.com
5	03OCT2017	26	6	7	bbc.co
6	03OCT2017	30	9	9	bbcimg.co
7	03OCT2017	30	6	11	bbc.co
8	03OCT2017	94	6	15	bbc.co
9	03OCT2017	30	6	21	bbc.co
10	03OCT2017	49	6	22	bbc.co
11	03OCT2017	75	7	24	bbci.co
12	03OCT2017	30	6	29	bbc.co
13	03OCT2017	49	6	30	bbc.co
14	03OCT2017	77	7	32	bbci.co
15	03OCT2017	30	6	37	bbc.co
16	03OCT2017	49	6	38	bbc.co
17	03OCT2017	75	7	40	bbci.co

4.4.5 Corporation Name

Capturing proper names of any kind is difficult to do, even when employing very sophisticated natural language algorithms. Doing so effectively without that background knowledge is honestly not worth doing unless you have an incredibly narrow, simplistic context and data source. However, performing the extraction of corporate entity references is much more feasible, especially if we leverage hot lists of corporation names (with variations).

> **Hot list:** A list of words already known, which we want to extract.

So, that is the only form of proper names that we will discuss extracting here. You will see many cases where even corporation names are misidentified by the relatively simple baseline pattern I am using here; however, there is plenty of opportunity for improvement based on the specific context for application.

Corporation pattern: "/(\b[A-Z]\w+\s[A-Z]\w+(\s[A-Z]\w+)*\b)|(\b(\w+\s+)*\w+\s+\ucorp(oration)?\b|\uinc\.?\b|\uco\.?\b|LLC\b|Company\b)/o"

This initial pattern matches under one of two circumstances; either a series of two or more words, each beginning with a capital letter, or a series of words followed by an abbreviation or word denoting a company (e.g., LLC, Co., or Inc.).

Output 4.6: Sample Initial Corp Pattern Results

Obs	ExtractDate	position	length	line	found
1	03OCT2017	25	8	5	BBC News
2	03OCT2017	31	8	6	BBC News
3	03OCT2017	20	8	10	BBC News
4	03OCT2017	44	32	15	British Broadcasting Corporation
5	03OCT2017	35	23	20	The European Commission
6	03OCT2017	35	13	36	But Instagram
7	03OCT2017	29	10	43	Apple Macs
8	03OCT2017	29	9	59	Elon Musk
9	03OCT2017	104	8	60	New York
10	03OCT2017	29	9	67	HSBC Beta
11	03OCT2017	29	12	75	Belle Gibson
12	03OCT2017	35	13	76	An Australian
13	03OCT2017	62	12	76	Belle Gibson
14	03OCT2017	29	12	83	GoPro Fusion
15	03OCT2017	35	10	84	The Fusion
16	03OCT2017	68	7	131	Face ID
17	03OCT2017	35	10	132	Even Apple
18	03OCT2017	29	28	139	Facebook CEO Mark Zuckerberg
19	03OCT2017	35	15	140	Mark Zuckerberg

It appears that my pattern creates too many matches to other entity types (and unusual phrases due to unexpected capitalization). What could be done to fix this? Well, I can make an initial attempt to fix it by simply restricting the pattern to include only explicit corporation identifiers in the name. For example: XYZ Inc., ABC Corp, and MyCo LLC are fictional company names with some of the corporate identifiers in the name.

Using the below pattern, you can see from my results in Output 4.7 that I'm missing a lot of potentially useful information. So, what am I supposed to do? It depends. As I continue to say, your context and business goals play a critical role in deciding how much time and effort you want to commit to finding the middle ground between these two extremes. For example, how important would it be for you to capture references to companies of interest that exclude the full company name (e.g., SAS instead of SAS Institute Inc.)? That outcome is in the middle ground, and it is very difficult without a hot list of company names and nicknames.

Updated corporation pattern:
"/(\b(\w+\s+)*\w+\s+\ucorp(oration)?\b|\uinc\.?\b|\uco\.?\b|LLC\b|Company\b)/o"

Output 4.7: Sample Initial Corp Pattern Results

Obs	ExtractDate	position	length	line	found
1	03OCT2017	44	32	15	British Broadcasting Corporation

4.5 Putting Them Together

Now that we have discussed the details of scraping text data, best practices for parsing it, and basic patterns to extract entity references, it's time to put it all together. I have constructed the code below to take the data now available on the file system, regardless of its origins, and process it for patterns of interest to us. I ran the code below using the Corp_Pattern variable in order to grab corporation entity references from our folder of text files. Knowing what is in the source folder, I'm curious how well this will perform, given the diversity of sample files.

The below macro picks up where I left off earlier, by expecting the results of that code to be present at run time. In order to make my code dynamically respond to the number of files, their location, and their names, I chose to build a macro to loop through each of the identified files and process them individually. The results of each loop though are printed using PROC PRINT.

This macro definition is relatively dense; so, I'm going to walk you through it in sufficient detail via the numbered items below.

Note: I have no intentions of making this an entire book on macro programming, but also don't anticipate you have a solid background in it. So, I will describe some elements without a lot of detail, but sufficient references for you to review on your own.

```
%macro parsing;
%do i=1 %to &filecount;   ❶
data parsing_result_&i;
infile %scan(&filelist,&i,%STR( )) length=linelen lrecl=500 pad;
varlen=linelen-0;          ❷
input source_text $varying500. varlen;
length sourcetable $50;
sourcetable=%scan(&filenames,&i,%STR( ));   ❸

Corp_Pattern = "/(\b[A-Z]\w+\s[A-Z]\w+(\s[A-
Z]\w+)*\b)|(\b(\w+\s+)*\w+\s+\ucorp(oration)?\b|\uinc\.?\b|\uco\.?\b|LLC\b|Compan
y\b)/o";
Website_Pattern = "/\b\w+\.(com|co|org|edu|gov|net)\b/o";
SSN_Pattern = "/\b\d{3}\s*-\s*\d{2}\s*-\s*\d{4}\b/o";
Phone_Pattern = "/(\+?\d\s*(-|\.|\()))?\s*?\d{3}\s*(-|\.|\)))\s*\d{3}\s*(-
|\.)\s*\d{4}/o";
DOB_Pattern = "/\d{1,2}\s*\/\s*\d{1,2}\s*\/\s*\d{4}/o";
Addr_Pattern = "/\s+(\w+(\s\w+)*\s\w+),?\s+(\w+\s*\w+),?\s+(\w+),?\s+((\d{5}\s*-
\s*\d{4})|\d{5})/o";

pattern_ID = PRXPARSE(Corp_Pattern);
start = 1;
stop = length(source_text);
CALL PRXNEXT(pattern_ID, start, stop, source_text, position, length);
   do while (position > 0);
      line=_N_;
      found = substr(source_text, position, length);
      put "Line:" _N_ found= position= length= ;
      output;
      CALL PRXNEXT(pattern_ID, start, stop, source_text, position, length);
   retain source_text start stop position length found;
   end;

keep sourcetable line position length found;
```

```
run;

proc print data=parsing_result_&i;  ❹
run;
%end;

%mend parsing;    ❺
%parsing;
```

❶ I open the macro definition for the macro named Parsing using the %MACRO statement. I then begin the macro DO loop that loops for the number of files identified in the prior step, and written to the macro variable Filecount. Next, I establish a data set name that is indexed against the file number.

❷ The INFILE statement uses the %SCAN function to parse the macro variable Filelist to insert the fully delineated file name for proper submission to SAS.

❸ The variable Sourcetable is created by using the %SCAN function again to parse the macro variable Filenames in order to get just the file being processed. I did this in order to track the source file inside of the data set. This will become important later when trying to identify source references.

❹ The PROC PRINT is referencing the indexed data sets generated above in order to print out the results. It is needed at this stage only to test whether the code is working correctly.

❺ And finally, the macro definition ends using the %MEND statement. And I immediately call the macro using %PARSING, the name assigned at the opening %MACRO statement.

4.6 Summary

I went through the entire process of entity extraction in this chapter—planning the strategy based on business needs, establishing methods for scraping a variety of text sources, creating baseline extraction patterns, and finally putting it all together using macros with notions of large-scale automation. I hope it has given you a solid understanding of how to get started with entity extraction for your next project. The code provided should act as a framework, while the patterns relevant to your application can be filled in to meet your specific needs.

As you can see, we have a tremendous amount of flexibility in how we execute this phase of work; so, it is important to continue referring back to the project goals and constraints to guide you. You also have to think about the technical needs of the future steps to ensure you have all the information necessary to achieve the business goals.

In the next chapter, I will explore the powerful ETL capabilities of the SAS language. What we have generated so far would qualify as "staging tables" with entity references from sources. The next chapter will give you the tools to clean up your structure data sources (yet to be discussed here), as well as the staging tables I just discussed.

---

[1]SAS Institute Inc. "HTTP Procedure," *Base SAS® Procedures Guide, Seventh Edition,* http://documentation.sas.com/?docsetId=proc&docsetTarget=n0bdg5vmrpyi7jn1pbgbje2atoov.htm&docsetVersion=9.4&locale=en (accessed August 29, 2018).

Chapter 5: Extract, Transform, Load

5.1 Introduction

Extract, Transform, and Load (ETL) is the process by which all source data is manipulated for downstream use in storage and analysis systems. The source can be raw data streams, flat files, staging tables, or production database tables. This stage of work is critical to ensuring that clean, useable data is entering the analytical phase of our process. I have attempted to break out the key elements of sound ETL processing to make clear the various things that you might need to actually do on the source data. Keep in mind that this is not an exhaustive list, but instead a list of techniques over and above your basic DATA step processing techniques to support ERA.

Figure 5.1: ERA Flow with ETL Focus

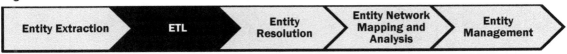

There are a number of standard techniques to be applied to your data at this stage in order to prepare it for analysis. However, as a best practice, you need to understand a bit about the data sources before applying any ETL techniques to them.

5.2 Examining Data

"Examining" is not an official ETL task, but just a smart thing to do before starting any ETL activities. You should always know what is actually present in your data sources of choice rather than acting on the information provided by a data dictionary or subject matter expert—although those sources are incredibly helpful guides.

Therefore, I am going to walk through a few procedures that will assist you in doing just that. The first will be PROC CONTENTS, which allows you to get metadata on libraries and data sets. This information can prove invaluable when making decisions about how to handle them. The second is PROC FREQ, which ensures you can examine (among other things) the values contained within data sets or columns.

5.2.1 PROC CONTENTS

The CONTENTS procedure is the best means of establishing your basic understanding of a data set that is already registered in SAS metadata—either because it is a SAS data set or registered in the metadata server via a SAS/ACCESS engine. I will not explore all of the options with PROC CONTENTS here, just what we need in order to read the basic metadata information that enables sound decision-making.

I'm using the CARS data set from the SASHELP library in my example below. The code below is a simple implementation of PROC CONTENTS, using the default settings.

```
proc contents data=sashelp.cars;
run;
```

As you can see in Output 5.1 below, I get some useful metadata to include the encoding scheme by just using the default setting the above code. It also shows the engine used for processing, which is V9 in my case. However, if you are running on SAS Viya, the engine listed will be "CAS" for the CAS engine that underpins SAS Viya.

Output 5.1: PROC CONTENTS Metadata Output

The CONTENTS Procedure			
Data Set Name	SASHELP.CARS	Observations	428
Member Type	DATA	Variables	15
Engine	V9	Indexes	0
Created	11/10/2016 00:57:44	Observation Length	152
Last Modified	11/10/2016 00:57:44	Deleted Observations	0
Protection		Compressed	NO
Data Set Type		Sorted	YES
Label	2004 Car Data		
Data Representation	WINDOWS_64		
Encoding	us-ascii ASCII (ANSI)		

Important things to note in the metadata shown in Output 5.1 above include the number of observations, variables, and indexes. These pieces of information allow you to know right away if a data set contains the number of these things that you expected. Also, the created and modified dates listed help you to know the age of your data, which is especially important if you don't have date variables in the data set.

In Output 5.2 below, you can see that PROC CONTENTS is providing information about the individual variables as well, including Type, Length, Format, and Label. Such information is critical when making decisions about what ETL steps need to be taken on a data set.

Output 5.2: Column Metadata

#	Variable	Type	Len	Format	Label
	Alphabetic List of Variables and Attributes				
9	Cylinders	Num	8		
5	DriveTrain	Char	5		
8	EngineSize	Num	8		Engine Size (L)
10	Horsepower	Num	8		
7	Invoice	Num	8	DOLLAR8.	
15	Length	Num	8		Length (IN)
11	MPG_City	Num	8		MPG (City)
12	MPG_Highway	Num	8		MPG (Highway)
6	MSRP	Num	8	DOLLAR8.	
1	Make	Char	13		
2	Model	Char	40		
4	Origin	Char	6		
3	Type	Char	8		
13	Weight	Num	8		Weight (LBS)
14	Wheelbase	Num	8		Wheelbase (IN)

5.2.2 PROC FREQ

Once you have the basic facts about a data set, you can then use the FREQ procedure to get a deeper understanding of the data columns in your source. You have to be careful here as the FREQ procedure will, by default, provide frequency information for numeric variables as well. Continuing with the SASHELP.CARS example in the code below, you can see the preferred use of PROC FREQ with Output 5.3, categorical variables.

```
Proc freq data=sashelp.cars;
run;
```

Output 5.3: Sample of the CARS Categorical Variables

Type	Frequency	Percent	Cumulative Frequency	Cumulative Percent
Hybrid	3	0.70	3	0.70
SUV	60	14.02	63	14.72
Sedan	262	61.21	325	75.93
Sports	49	11.45	374	87.38
Truck	24	5.61	398	92.99
Wagon	30	7.01	428	100.00

Origin	Frequency	Percent	Cumulative Frequency	Cumulative Percent
Asia	158	36.92	158	36.92
Europe	123	28.74	281	65.65
USA	147	34.35	428	100.00

DriveTrain	Frequency	Percent	Cumulative Frequency	Cumulative Percent
All	92	21.50	92	21.50
Front	226	52.80	318	74.30
Rear	110	25.70	428	100.00

Below, in Output 5.4, is an example of what to avoid with the FREQ procedure. Depending on the data set with which you are dealing, you may fill up your memory, or just create a process that takes far too long to run. You will know in advance if any nightmare scenarios are likely, based on the PROC CONTENTS results discussed above. It is clearly not a concern in the CARS situation as there are only 428 observations in the entire data set.

Output 5.4: Sample of CARS MSRP Frequencies

MSRP	Frequency	Percent	Cumulative Frequency	Cumulative Percent
$10,280	1	0.23	1	0.23
$10,539	1	0.23	2	0.47
$10,760	1	0.23	3	0.70
$10,995	1	0.23	4	0.93
$11,155	1	0.23	5	1.17
$11,290	1	0.23	6	1.40
$11,560	1	0.23	7	1.64
$11,690	1	0.23	8	1.87
$11,839	1	0.23	9	2.10
$11,905	1	0.23	10	2.34
$11,939	1	0.23	11	2.57
$12,269	1	0.23	12	2.80
$12,360	1	0.23	13	3.04
$12,585	1	0.23	14	3.27
$12,740	1	0.23	15	3.50
$12,800	1	0.23	16	3.74
$12,884	1	0.23	17	3.97

Simply having a numeric variable doesn't mean you will have many unique values to handle, even with large data sets. For example, a binary variable (i.e., 0 or 1 only) could be numeric. However, for data sets with many observations, it is safest to examine numeric variables with the PROC MEANS statement, as discussed next.

5.2.3 PROC MEANS

The MEANS procedure provides useful information about numeric variables, enabling you to make decisions about the best way to handle them. Below is the code to run the MEANS procedure with default settings for the CARS data set. You can see the default columns in Output 5.5 below as well.

```
proc means data=sashelp.cars;
run;
```

Output 5.5: Default PROC MEANS Results

Variable	Label	N	Mean	Std Dev	Minimum	Maximum
MSRP		428	32774.86	19431.72	10280.00	192465.00
Invoice		428	30014.70	17642.12	9875.00	173560.00
EngineSize	Engine Size (L)	428	3.1967290	1.1085947	1.3000000	8.3000000
Cylinders		426	5.8075117	1.5584426	3.0000000	12.0000000
Horsepower		428	215.8855140	71.8360316	73.0000000	500.0000000
MPG_City	MPG (City)	428	20.0607477	5.2382176	10.0000000	60.0000000
MPG_Highway	MPG (Highway)	428	26.8434579	5.7412007	12.0000000	66.0000000
Weight	Weight (LBS)	428	3577.95	758.9832146	1850.00	7190.00
Wheelbase	Wheelbase (IN)	428	108.1542056	8.3118130	89.0000000	144.0000000
Length	Length (IN)	428	186.3621495	14.3579913	143.0000000	238.0000000

This default information can be very helpful; however, using additional options for PROC MEANS can prove more insightful, enabling your decision-making for what to do with each column in subsequent ETL steps.

So, I've updated the code to show a few different statistics about the CARS data set.

```
proc means data=sashelp.cars n nmiss min q1 mean q3 max std;
run;
```

Using the listed options, I'm getting the number of observations, number of missing values, minimum value, first quartile, mean, third quartile, maximum value, and standard deviation for each column. There are additional options that you can explore in the documentation, but these will provide the information that we need for numerical columns the majority of the time. Output 5.6 shows the results for the CARS data set.

Output 5.6: Custom PROC MEANS Output

Variable	Label	N	N Miss	Minimum	Lower Quartile	Mean	Upper Quartile	Maximum	Std Dev
MSRP		428	0	10280.00	20329.50	32774.86	39215.00	192465.00	19431.72
Invoice		428	0	9875.00	18851.00	30014.70	35732.50	173560.00	17642.12
EngineSize	Engine Size (L)	428	0	1.3000000	2.3500000	3.1967290	3.9000000	8.3000000	1.1085947
Cylinders		426	2	3.0000000	4.0000000	5.8075117	6.0000000	12.0000000	1.5584426
Horsepower		428	0	73.0000000	165.0000000	215.8855140	255.0000000	500.0000000	71.8360316
MPG_City	MPG (City)	428	0	10.0000000	17.0000000	20.0607477	21.5000000	60.0000000	5.2382176
MPG_Highway	MPG (Highway)	428	0	12.0000000	24.0000000	26.8434579	29.0000000	66.0000000	5.7412007
Weight	Weight (LBS)	428	0	1850.00	3103.00	3577.95	3978.50	7190.00	758.9832146
Wheelbase	Wheelbase (IN)	428	0	89.0000000	103.0000000	108.1542056	112.0000000	144.0000000	8.3118130
Length	Length (IN)	428	0	143.0000000	178.0000000	186.3621495	194.0000000	238.0000000	14.3579913

While some additional exploratory data analysis may be desired, depending on the nature of your source data, the brief discussion above should provide you enough information to make sound decisions about the ETL steps that you need to take for entity resolution.

5.3 Encoding Translation

The SAS Platform has been around in some form since the 1970s, when mainframes were the primary form of high-powered computing. As technology has changed, the SAS language has adapted without losing the ability to communicate with these legacy systems. To facilitate smooth transitioning through an ever-changing technology landscape, the platform has the ability to translate or convert the encoding of data formats between various storage platforms.

A great example of this goes all the way back to mainframe computers. Without getting bogged down in the details, EBCDIC is a compact way many mainframes used to store information. Memory was far more expensive and precious than it is in modern systems; so, manufacturers of these systems had to create an encoding scheme for data that was very compact. However, its compactness does not lend itself to being readable by humans. Modern operating systems use ASCII encoding, which ensures you can read the data stored by them. I have chosen EBCDIC specifically because legacy mainframes are still floating around many federal and state government agencies, as well as some large financial institutions. So, it is practical to be aware of this format, and where you might see it pop up.

Note: Modern websites use UTF-8 or UTF-16 data encoding. The steps for those data sources will be handled using the same method I'm describing here; so, please bear with me.

Below is a simple example to demonstrate how we can actually encode data in EBCDIC format with the DATA step option ENCODING. In this step, I create a data set named Encode with a single record containing the variables x and abc123 (not creative names, but they demonstrate the idea). The variables have the values, 1 and abc123, I typed on my ASCII-base OS (Windows). The ENCODING= option is used to define my data set encoding as EBCDIC, meaning that is the encoding scheme SAS will use to write the data set to the destination library. This definition will be applied to the specified data set only, regardless of the encoding of other data sets in the associated library.

```
data work.encode (encoding="ebcdic");
x=1;
abc123 = 'abc123';
run;
```

Using what was just discussed above about PROC CONTENTS, you can check the encoding defined on the above data set, ENCODE.

```
proc contents data=work.encode;
run;
```

See the result of running the CONTENTS procedure, and notice that the row Encoding below has a value of ebcdic1047 Western (EBCDIC). There are different types of EBCDIC encoding, and this happens to be the default in SAS, unless otherwise specified.

Output 5.7: PROC CONTENTS Results

Data Set Name	WORK.ENCODE	Observations	1
Member Type	DATA	Variables	2
Engine	V9	Indexes	0
Created	12/06/2017 22:38:29	Observation Length	16
Last Modified	12/06/2017 22:38:29	Deleted Observations	0
Protection		Compressed	NO
Data Set Type		Sorted	NO
Label			
Data Representation	WINDOWS_64		
Encoding	ebcdic1047 Western (EBCDIC)		

I have verified that the data set ENCODE is encoded as EBCDIC. But I want to further demonstrate the impact on consumers of your data. To help demonstrate the effect this has on the data set, I have two PROC PRINT statements below. First note that once a data set encoding has been established, SAS automatically translates it to the encoding of the SAS host environment when reading into memory. If I were to print the data set without stating the encoding, it would just print the result of the conversion SAS did for me. So, I am explicitly identifying the encoding in my PROC PRINT statements below to make it clear that I am dealing with data that is encoded as EBCDIC. The first PRINT procedure uses the option encoding="ebcdic", while the second uses encoding="ascii".

```
proc print data=work.encode (encoding="ebcdic");
run;

proc print data=work.encode (encoding="ascii");
run;
```

Note the difference in output below. You can see this EBCDIC encoded data set returns the expected output when that same encoding option is used by PROC PRINT. However, when I use encoding=ASCII, I get unreadable output (also known as "garbage").

Output 5.8: PROC PRINT Output with EBCDIC and ASCII

Obs	x	abc123
1	1	abc123

Obs	§	‚ƒñòó
1	1	‚ƒñòó

Now that you've seen how to write a different encoding scheme with the DATA step ENCODING option, it's time to see how to *read* a data set with a different encoding. This is again quite straightforward.

I use the data set created above as my source for the below example. In the DATA statement, I define a new data set Work.encode2 with an ASCII encoding. And using the SET statement, I identify the existing data set Work.encode, along with its encoding scheme of EBCDIC.

```
data work.encode2 (encoding="ascii");
set work.encode (encoding="ebcdic");
run;
```

I repeat the PROC PRINT statements from the first example below, with the new data set (Encode2). Notice how the results from Output 5.8 are flipped in Output 5.9 below. This demonstrates how the encoding is now ASCII rather than EBCDIC as it was before.

```
proc print data=work.encode2 (encoding="ebcdic");
run;
proc print data=work.encode2 (encoding="ascii");
run;
```

Output 5.9: PROC PRINT Output with EBCDIC and ASCII

Obs	ï	/ÂÄ"•
1	1	/ÂÄ"•

Obs	x	abc123
1	1	abc123

As you can see, encoding conversion of data can be simple via the SAS DATA step, but critically important to get right. Otherwise, you could have a mess to clean up later. And again note that this process works with whatever encoding schemes you need to translate between—SAS can be the nexus point for many different data environments.

5.4 Conversion

It is often necessary to convert data formats when pulling from source tables in order to standardize the formats across multiple data sets, or to enable a later analysis. This is distinct from encoding translation in that all the source data columns are using the same encoding scheme (e.g., ASCII, EBCDIC), and will need to be done after the encoding is verified as accurate.

5.4.1 Hexadecimal to Decimal

Below is a short example of how you can convert hexadecimal numbers to decimal numbers. As you can see in the code below, I have DATALINES values with hexadecimal numbers.

Note: Hexadecimal is base 16 rather than base 10, which means that the numbers go 0 to F for a single "digit" instead of 0 to 9. Please see Appendix A for relevant POSIX metacharacters.

```
data hex2dec_example;
input nums Hex2.;
datalines;
00
02
99
FF
AF
;
run;

proc print data=work.hex2dec_example;
run;
```

I'm using informat HEX2 in the above code to tell SAS that the DATALINES values are two-digit hexadecimal numbers. As you can see in Output 5.10, the PRINT procedure displays the successful conversion to decimal numbers for the variable NUMS. Decimal is the default numerical data storage format for SAS; so, the informat conversion from hexadecimal is all I need to provide.

Output 5.10: Decimal Representation of Hexadecimal Numbers

Obs	nums
1	0
2	2
3	153
4	255
5	175

5.4.2 Working with Dates

It is also commonly necessary to convert date formats. Whether you are converting from European to American date formats, or four-digit to two-digit years, SAS has an informat option for the task at hand. I will just cover a couple of options below, but there are *so many different date formats that you may encounter*. It all starts with analyzing your data ahead of time to understand what format you are receiving from the source.

In the below code, I have created data lines that are likely something that you will encounter; unformatted numerical representations of European dates, one with 2-digit years and one with 4-digit years. I have only two lines to demonstrate the concept. Note the DDMMYY informat is used for the column Euro while DDMMYY10 is used for the column FourDigit. An unformatted date with the day-month-year order is often called a European date because that is the standard order in Europe, while the month-day-year order is standard in the United States.

```
data DateConversion;
input Euro ddmmyy. FourDigit ddmmyy10.;
datalines;
010118 01012018
120117 12012017
;
run;
```

When you apply the correct informat, SAS stores the variables as internal date numbers, which allows you to then display the date using any preferred format. I use PROC PRINT below to display the date in the U.S. standard format of MMDDYY.

```
proc print data=work.DateConversion;
format euro mmddyy8. fourdigit mmddyy10.;
run;
```

Output 5.11: US Date Format

Obs	Euro	FourDigit
1	01/01/18	01/01/2018
2	01/12/17	01/12/2017

Now, just to emphasize the point about dates, I have created PROC PRINT output below that uses a different display format for the same internally stored date numbers.

```
proc print data=work.DateConversion;
format euro date7. fourdigit date10.;
run;
```

Output 5.12: Abbreviated Month Date Format

Obs	Euro	FourDigit
1	01JAN18	01JAN2018
2	12JAN17	12JAN2017

I want to emphasize that this is a very small taste of the litany of possible date formats out there. Rather than expending many pages trying to show you many, but not nearly all, of the other formats out there, I have chosen to provide a couple of fairly common tasks here. I would encourage you to research other formats in the SAS documentation.[1]

5.5 Standardization

Data standardization takes on different meanings, depending on the data type in the source field that you are trying to standardize—numeric or character. For numeric fields, we would normally use PROC STDIZE, or a similar method, to perform a statistical standardization of the data. However, in the case of character values, it means we will change the character field values to be consistent for every record. Since the goal of this activity is to prepare data for entity resolution activities, I will focus on character standardization as that will be what we need to improve match quality.

Standardizing a character field is very straightforward, but there unfortunately isn't a simple procedure to do the heavy lifting for you. You must assess all possible values in a field, and map them to the same standard character representations. This step is most often taken to ensure consistency across data sources; however, it is also helpful to perform on large data sets as a means of reducing the data storage needed. Every ASCII character is a byte of storage; so, trimming a few characters off of a column length can turn into huge memory savings on very large data sets.

Note: You save memory only if the new SAS data column is the smaller format! Be sure to change the format definition on your resulting data set.

For example, yes, y, and 1 can all be mapped to Y, while all versions of no can be mapped to N; this ensures a standardized binary representation for the concept of yes/no or true/false while using less memory. You might think I've gotten something backwards by deferring to characters rather than numbers for creating binary variables. However, keep in mind that SAS requires a minimum of 2 bytes to store a numeric (the minimum can vary by operating system) on ASCII systems like Windows, while it requires a minimum of only 1 byte to store a single character.[2] So, if space is truly at a premium, a single-byte character is what you will want to do in SAS.

I have taken this approach for the simple example below by using a SELECT statement to remove the noise in my source data.

```
/*Character Standardization*/
data charSTD_example;
input answer $;

select(answer);
    when('yes','y','1') STDans='Y';
    when('no','No','n','0') STDans='N';
    otherwise;
    end;
datalines;
yes
y
1
no
n
No
run;

proc print data=work.charSTD_example;
run;
```

The above code creates the output shown in Output 5.13 below. As you can see, all the various values for the column that is called answer are standardized to the binary Y/N I want. Could this be done through other means? Of course, but I have attempted to provide a method that is intuitive and accessible via Base SAS code.

Output 5.13: Standardization Results

Obs	answer	STDans
1	yes	Y
2	y	Y
3	1	Y
4	no	N
5	n	N
6	No	N

5.6 Binning

Numeric variables are clearly very valuable in their native form. However, depending on the intended use, you may want to break them into categories instead. Doing so can make them more valuable for data segmentation, predictive modeling, and entity resolution.

Binning for entity resolution helps us ensure that small discrepancies in numeric values don't prevent the matching of two entity references (discussed more in the next chapter). You have to use your best judgment regarding which variables are the ideal candidates for binning.

Example: Suppose you have multiple real estate listings for the same house, with slightly different numerical details such as price. By binning the price for that address, you will have an improved chance of getting a multi-factor match during the actual matching process performed during the ER phase of work.

I will demonstrate a couple binning strategies with the HPBIN procedure below. Your business problem will drive the best choice for you.

HPBIN stands for High-Performance Binning; it is a procedure that is flexible enough to run in a high-performance environment across numerous nodes, or on your desktop. I've chosen a sample data set from the SASHELP library, which is available with all SAS products. It provides some simple numerical variables to bin using a couple of different approaches, discussed below.

PROC HPBIN provides a great deal of flexibility for how you want to bin your numeric variables. As I mentioned, your particular end-use will drive the best binning strategy for you, but I am covering two very common methods below: quantile and bucket.

5.6.1 Quantile Binning

First is a quantile-based approach, where you want the number of records to be roughly equal in each bin. You can determine how many quantiles you want to build after examining your data (as we did in Section 5.2 above), but I'm going to stick with 4 (quartiles) for my example below.

```
proc hpbin data=sashelp.baseball numbin=4 pseudo_quantile computequantile;
input nHits;
run;
```

By setting my NUMBIN equal to 4, I am telling HPBIN to create 4 bins. I then tell HPBIN to do that with the "pseudo_quantile" method. This creates 4 quantile bins or quartiles. You can see the results in the output below.

Output 5.14: PROC HPBIN Settings

Performance Information	
Execution Mode	Single-Machine
Number of Threads	4

Data Access Information			
Data	Engine	Role	Path
SASHELP.BASEBALL	V9	Input	On Client

Binning Information	
Method	Pseudo-Quantile Binning
Number of Bins Specified	4
Number of Variables	1

Output 5.14 provides a fair amount of detail about the settings. But take particular notice of the Method and Number of Bins Specified. You see the method defined as Pseudo-Quantile Binning, which creates the bins using a quantile method, resulting in nearly equal sized bins (by frequency). You can see in Output 5.15 below that the bin ranges match the quartile values in the Quantiles and Extremes table also shown in Output 5.15 below.

Output 5.15: Results of PROC HPBIN

Mapping					
Variable	Binned Variable	Range		Frequency	Proportion
nHits	BIN_nHits	nHits < 68.0116		84	0.26086957
		68.0116 <= nHits < 98.0059		77	0.23913043
		98.0059 <= nHits < 138.019		82	0.25465839
		138.019 <= nHits		79	0.24534161

Quantiles and Extremes		
Variable	Quantile Level	Quantile
nHits	Max	238
	.99	211
	.95	174
	.90	163
	.75 Q3	138
	.50 Median	98
	.25 Q1	68
	.10	49
	.05	41
	.01	34
	Min	31

5.6.2 Bucket Binning

The second approach is to create bins using equal-length numerical ranges of the source variable. So, rather than trying to ensure that the number records are roughly equal, as we did above, the number of records can vary wildly; only the numerical range that defines each bin is equal in size. So, using the same data set as before, I have created equal length bins using the defaults for PROC HPBIN below.

```
proc hpbin data=sashelp.baseball;
input nHits;
run;
```

As you can see in Output 5.16, the bucket binning method creates 16 bins spanning the same length of the nHits range. The resulting bins can have very different frequencies, as you can see below.

Output 5.16: Bucket Binning Results

Binning Information	
Method	Bucket Binning
Number of Bins Specified	16
Number of Variables	1

		Mapping		
Variable	Binned Variable	Range	Frequency	Proportion
nHits	BIN_nHits	nHits < 43.9375	22	0.06832298
		43.9375 <= nHits < 56.875	34	0.10559006
		56.875 <= nHits < 69.8125	32	0.09937888
		69.8125 <= nHits < 82.75	35	0.10869565
		82.75 <= nHits < 95.6875	31	0.09627329
		95.6875 <= nHits < 108.625	28	0.08695652
		108.625 <= nHits < 121.5625	29	0.09006211
		121.5625 <= nHits < 134.5	22	0.06832298
		134.5 <= nHits < 147.4375	31	0.09627329
		147.4375 <= nHits < 160.375	22	0.06832298
		160.375 <= nHits < 173.3125	18	0.05590062
		173.3125 <= nHits < 186.25	8	0.02484472
		186.25 <= nHits < 199.1875	2	0.00621118
		199.1875 <= nHits < 212.125	5	0.01552795
		212.125 <= nHits < 225.0625	2	0.00621118
		225.0625 <= nHits	1	0.00310559

The two binning methods discussed above demonstrate very different results, with different uses. For variables with very sparse data and wide ranges, a quantile method would be better for grouping them. However, for data that has a small range or is quite dense, a bucket binning approach may be more effective.

5.7 Summary

As mentioned in the outset of this chapter, I hope the information discussed provides you additional tools for wrangling your data sources. While there is nothing magical about the techniques discussed, I think they will enable you to tackle the significant task of preparing your various data sources for integration and analysis.

Data cleaning and preparation via a robust set of ETL processes will provide you results that are ready for fusion and entity resolution as discussed in the next chapter.

---

[1] SAS Institute Inc. "Formats by Category," *SAS® 9.4 Formats and Informats: Reference,* http://support.sas.com/documentation/cdl/en/leforinforref/64790/HTML/default/viewer.htm#n0p2fmevfgj4 70n17h4k9f27qjag.htm (accessed August 29, 2018).

[2] SAS Institute Inc. "Ways to Create Variables," *SAS® 9.4 Language Reference: Concepts, Sixth Edition,* http://go.documentation.sas.com/?docsetId=lrcon&docsetTarget=n0bbin3txgxlown1v2v5d8qbc9vq.htm&d ocsetVersion=9.4&locale=en (accessed August 29, 2018).

Chapter 6: Entity Resolution

6.1 Introduction

There are a number of robust methods for performing entity resolution efficiently at scale. But despite their diversity, they all fall into two basic families: exact and fuzzy.

Figure 6.1: ERA Flow with Entity Resolution Focus

6.1.1 Exact Matching

Exact matching performs an exact comparison of each entity reference attribute, and makes a "match" determination based on how many attributes are the same. A previously defined rule for similarity—a threshold for percentage of attributes covered—provides the decision mechanism. As I mentioned during the introduction to ERA in Chapter 3, this threshold is informed by the business context. The need to get a match exactly correct is going to vary by the application as the error tolerance will be very different across domains.

In contexts where entity false positives are of chief concern, a higher similarity percentage would be preferred, while contexts most concerned with false negatives would allow a much lower similarity percentage. So, if you set the match tolerance such that you must get a 100% similarity, there will be some number of entity references that will not be matched together—creating false negatives. However, if the tolerance is set very low, requiring a 50% match, then the false positives could be quite high. Your context would drive the determination as to what tradeoffs you are willing to make. And detailed analysis of sample data will help you quantify these tradeoffs.

Example: The table below shows two records with similar, but not identical, information. These records could be matched, depending on the matching tolerance—which is driven by the eventual business need and associated risk tolerance. If we simply utilize a raw percentage of match, then we have six out of eight fields with exactly the same data (75%). Now, in many cases, certain values are far more important than the others. So, the business may decide to be concerned only about the percentage of match between the first four data columns (coincidentally, for this example, it is also 75%). However, if the business requirements set the threshold at 87.5% (7 out of 8), these records would not meet the criteria under either approach (comparing all 8 or comparing just the first 4).

Table 6.1: Similar Entity Reference Records

First Name	Last Name	DOB	Gender	Address	City	State	Zip
Bob	Smith	2/5/1967	M	123 Fourth Street	Fairfax	VA	22030
Robert	Smith	2/5/1967	M	123 4th St.	Fairfax	VA	22030

In addition to expert opinion for determining the most important variables for building unique matches, you can infer a weighting through careful analysis. You can sample the available data, build matches with it, and then analyze the frequency of match by each variable. If a single variable created unique matches (e.g., by SSN), it would clearly be important. However, variables that create many non-unique matches would be less important. Then looking at the two, three, etc., variable combinations that create unique matches will enable you to rank or group those that will be required for a "match" to be accepted for the whole population.

6.1.2 Fuzzy Matching

A fuzzy matching approach measures similarity between each entity reference element being compared, computes the overall entity reference similarity, and determines a match based on the defined threshold. The similarity of each attribute is derived based on the kind of data that it contains. If it is a numeric field, then a numerical distance is calculated. On the other hand, if it is a text field, then semantic or syntactic comparisons can be performed.

Semantic match: The meaning of the words or phrases used is the same.

Syntactic match: The arrangement of words or phrases used is the same.

Some of the advanced techniques, such as semantic matching, are beyond the scope of this book, but I will still explore a robust set of matching techniques in Section 6.3 using edit distance concepts—which is a kind of syntactic matching.

Example: Referring to the table in the previous example, the address line seems to simply be different ways of writing the same information. If I standardize the address information, then you see that the address line is the same for both records.

Table 6.2: Modified Entity Reference Records

First Name	Last Name	DOB	Gender	Address	City	State	Zip
Bob	Smith	2/5/1967	M	123 4th St.	Fairfax	VA	22030
Robert	Smith	2/5/1967	M	123 4th St.	Fairfax	VA	22030

- Now, while this sort of normalization should be performed at the database level during ETL, it is not always done properly. In addition, it is not always feasible to anticipate this level of normalization for such data when it is being pulled from multiple sources—often the case in real-world applications. Therefore, performing the matching operation of this information requires that a syntactic match be applied. The result generates a 100% match of the address lines, since they are just different ways to write the same information. This in turn changes the percentage of overall match to 7 out of 8 columns.

Achieving an 8 out of 8 match at this point would require a semantic match on the first name. This is where things become truly fuzzy; since there is no way to know for sure that "Bob" and "Robert" are references to the same person. This sort of determination is either asserted via an expert rule, or learned from a significant corpus of *culturally similar* data. The asserted or learned similarity would then be applied to the name field, and incorporated with the overall percentage of match. This new number, depending on the algorithm chosen for integration and the similarity value, would then be tested against the known match threshold.

Explaining the last step described in the above example is beyond the scope of this book. The domain-specific knowledge, corpus of training data, and detailed examples needed to properly cover the topic are beyond what I have the capacity to cover here. However, performing syntactic matches via edit distance functions in SAS is reasonable, and what I will do in this book. I will also cover a method for making names easier to match with edit distance functions available in Base SAS, mitigating this potential gap.

Examples such as the one above make the matching process seem easy. However, automating matching decisions at scale is the very essence of what makes ERA so difficult. I therefore want to emphasize the importance of consistent matching decisions via a framework like ERA—creating institutional buy-in before rollout across any production systems is critical to ensure effective implementation.

In addition, ERA provides a decision framework for determining error tolerance for the particular application.

6.1.3 Error Handling

In the vast majority of entity resolution applications, Type-1 error (false positive) is far worse than Type-2 error (false negative).

Type-1 error in this context means that two entities are equivalent when they in fact are not, which among other issues could lead to the mishandling of Personally Identifiable Information (PII).

Example: An insurance company merges the records for two individuals erroneously, thus corrupting the medical history of two patients, possibly sharing their PII with the incorrect individual, and putting their health at risk.

By contrast, a Type-2 error would essentially cause duplicative entity references to be generated and maintained. While a frustration for both the organization and individual(s) affected, the risk to either party is lower than the Type-1 case.

Example: ABC Inc. (ABC) has duplicate customer records in their database, possibly causing an item to be shipped to an old address, or requiring support staff to clean out duplicate accounts; but the PII is kept safe. This represents an inconvenience, which is relatively easy to avoid or resolve.

While generally less dangerous, Type-2 error is also much harder to detect. Unless you have enough data to identify and fix such a problem, it will go completely unnoticed. In situations where Type-2 error has occurred, it will likely require customers to identify the issue rather than the company or agency.

The best strategy for reducing Type-2 errors, while reducing security risk, is to communicate directly with users. For example, by cross-referencing credit card information, ABC could internally identify accounts that are connected. Through email alerts with customers, ABC can provide them the opportunity to merge or close accounts with multiple PII verification steps involved. By removing these ghost accounts, the Type-2 error across the system is reduced, thus lowering the risk of old accounts getting hacked, and enabling ABC to better understand its real customer base.

6.2 Indexing

In order to effectively perform entity reference matching at scale, you have to properly index your source data tables. By this point in the ERA process, any data sources that you are working with have been ingested and staged as SAS data tables. So, I will focus this discussion on creating indexes on SAS data sets. By indexing said tables, we will be able to more efficiently perform matching of the entity references on a very large data set.

However, it is important to note that indexes are not always beneficial to processing performance. Once created, indexes are treated as part of the data set; as data set values change, the index is automatically updated. This results in overhead associated with building and maintaining indexes, as well as with their real-time use. So, you have to determine whether the size of your data will necessitate using indexes, which can depend on a number of factors.

Note: Indexes are ignored in the DATA step when a subsetting IF statement is used!

Here are a few guidelines to help you determine how and when to build and use indexes: [1]

- Do not create an index for a data set of less than three memory pages as the index overhead will ultimately prove counterproductive. As a general rule of thumb, you will see performance improvements for data sets larger than three data set memory pages. Use PROC CONTENTS to determine the number of pages for any data set.
- Data sets that change with high frequency will require updates to their indexes, creating increased resource demands, which may outweigh the overall benefit of the indexes. So, in general, avoid using indexes on data sets that change frequently, and run tests on data sets that you are unsure about prior to implementing an index at scale.
- Ensure that the index that you want to create is as discriminating as possible (i.e., it selects the fewest records per index value as possible). For example, indexing a data set by date on a data set that contains dozens or hundreds of records for each date is not at all discriminating. However, building an index on Last Name and Date of Birth will generate far fewer record matches per index value—more discriminating.

As with anything in SAS, there are multiple methods for creating and updating indexes on data sets. You can use PROC DATASETS, PROC SQL, or the INDEX= option in the DATA step. There are tradeoffs with each option, and you are welcome to explore all of them. However, the INDEX= option in a SAS DATA step is easy to understand and use, and fills the need for our purposes.

6.2.1 INDEX=

The INDEX= option is easy to implement, whether we want a simple index or composite index. I will demonstrate how to do both with INDEX= below.

Simple Index

A simple index is made up of only one variable in the target data set. The code below uses PII_DATA, which I will explore in more depth later in this chapter, to create PII_DATA2 with an index on the variable Last_Name.

```
data pii_data2 (index=(Last_Name));
set pii_data;
run;
```

When running the code above, assuming no errors, you won't be able to see any indication that an index was actually built. There is neither a reference in the SAS log, nor anything on the data set. So, running PROC CONTENTS on the resulting data set is the best way to know for sure that your index has been created. Running the following code, creates Output 6.1 below.

```
proc contents data=work.pii_data2;
run;
```

In addition to the index count now being equal to 1 in the PROC CONTENTS results, you will see the information in Output 6.1 at the bottom of your results. Immediately after the alphabetical list of variables, you see the alphabetical list of indexes. As expected, you see the index Last_Name. The data set we are using has only 25 rows; so, this is an index that uniquely identifies each record (i.e., it is highly discriminating).

Output 6.1: Simple Index in PROC CONTENTS Output

Alphabetic List of Variables and Attributes			
#	Variable	Type	Len
6	Address	Char	100
8	City	Char	100
5	DOB	Num	8
1	First_Name	Char	20
2	Last_Name	Char	20
3	Phone	Char	15
4	SSN	Char	11
9	State	Char	100
7	Street	Char	100
10	Zip	Char	100

Alphabetic List of Indexes and Attributes		
#	Index	# of Unique Values
1	Last_Name	25

Composite Index

Now, there will be situations where you need to use more than one variable to build an index. This is known as a composite index. Generally speaking, you will want to do this in order to ensure you can improve the discrimination of the index, ensuring that it is as efficient as possible.

For example, you will not always have unique entity reference identifiers such as Social Security numbers. In such a case, the composite, uniquely identifying index might be First Name, Last Name, and Date of Birth. The likelihood of those three variables uniquely identifying a person is quite high; however, this should always be verified in the data set with which you are working.

Below is an example of how I chose to build a composite index using the same data sets as before. Note that I ran the simple index DATA step and the below DATA step back-to-back without any problems. SAS merely overwrote the simple index created above with the composite index that you see below.

```
data pii_data2 (index=(comp_ind=(Last_Name DOB)));
set pii_data;
run;
proc contents data=work.pii_data2;
run;
```

The code is very similar to the simple index example above, but note that I have created a variable, COMP_IND, inside the first set of parentheses, and set it equal to a parenthetical list of variables, Last_Name and DOB. This creates a composite index called COMP_IND, as you can see in Output 6.2 below.

Output 6.2: Composite Index PROC CONTENTS Output

Alphabetic List of Variables and Attributes			
#	Variable	Type	Len
6	Address	Char	100
8	City	Char	100
5	DOB	Num	8
1	First_Name	Char	20
2	Last_Name	Char	20
3	Phone	Char	15
4	SSN	Char	11
9	State	Char	100
7	Street	Char	100
10	Zip	Char	100

Alphabetic List of Indexes and Attributes			
#	Index	# of Unique Values	Variables
1	comp_ind	25	Last_Name DOB

Unique and Nomiss Modifiers

There are two very helpful modifiers that you can use in the INDEX= data set option, Unique and Nomiss. The Unique modifier ensures that the index is built only if all the values for the variables that are used to build it are unique. And the Nomiss modifier ensures that rows with missing values are excluded from the index. Below is an example of how to use these modifiers when constructing an index.

```
data pii_data2 (index=(Last_Name /unique /nomiss));
set pii_data;
run;
proc contents data=work.pii_data2;
run;
```

Notice that the forward slash (/) must be applied before each modifier. Without the / present, SAS will interpret the modifier as an attempt to list a variable for your index, and throw an error. Below in Output 6.3 is the updated index information from PROC CONTENTS.

As you can see, Unique and NoMiss are both set to YES, indicating that these two modifiers are true for this index.

Output 6.3: Unique and NoMiss Index Modifiers

Alphabetic List of Indexes and Attributes				
#	Index	Unique Option	NoMiss Option	# of Unique Values
1	Last_Name	YES	YES	25

Keeping the guidelines at the beginning of this section in mind, you now have the basic tools to create or update indexes on data sets.

6.3 Matching

Exact matching is the most straightforward method for comparing entity references, while fuzzy matching is usually a bit more complicated. Both approaches can be performed through a variety of methods available in SAS. You can perform exact matching with traditional "equal" join criteria, while also making certain join criteria on other elements fuzzy with match-merges or PROC SQL. Again, as with many things in SAS, there are multiple methods for tackling the issue of joining or merging data sets. However, PROC SQL offers the most flexibility of all options; so, I will use that in Section 6.3.3 for bringing the matching concepts together as a complete example.

6.3.1 COMPGED and COMPLEV

Both exact and fuzzy matching can be performed in Base SAS with assistance from the COMPGED and COMPLEV functions by dialing match thresholds to a desired level. COMPGED provides the General Edit Distance (GED) between two string variables, while the COMPLEV function provides the Levenshtein Edit Distance (LEV)—a special case of the GED.

COMPLEV runs more efficiently than COMPGED, but the LEV result is not considered as useful in the context of text mining or fuzzy matching due to the simplicity of output.[2] Conversely, the output from COMPGED, while more computationally expensive, is more nuanced to support more refined measurement. Your ultimate needs will determine the best function.

I demonstrate output differences between the two methods in the example below. As you can see in the DATALINES section, I have chosen "balloon" as my word to match. I defined GED equal to the COMPGED result, and LEV as the COMPLEV. And the first row of data shows that the match between "balloon" and "balloon" is 0 for both GED and LEV. In other words, there are no edits needed to make the two strings equal. You can see the cost for each operation in Output 6.4 below, with both GED and LEV values provided. The list of operations is in the GED order of increasing cost.

```
data Edit_Dist;
   infile datalines dlm=',';
   input String1 : $char10. String2 : $char10. Operation $40.;
   GED = compged(string1, string2);
   LEV = complev(string1, string2);
   datalines;
balloon,balloon,match
ba lloon,balloon,blank
balloo,balloon,truncate
baalloon,balloon,double
ballon,balloon,single
balolon,balloon,swap
ball.oon,balloon,punctuation
balloons,balloon,append
balkloon,balloon,insert
blloon,balloon,delete
balloon,balloon,replace
blolon,balloon,swap+delete
balls,balloon,replace+truncate*2
balXtoon,balloon,replace+insert
blYloon,balloon,insert+delete
bkakloon,balloon,insert+replace
bllooX,balloon,delete+replace
kballoon,balloon,finsert
alloon,balloon,fdelete
kalloon,balloon,freplace
akloon,balloon,fdelete+replace
akloo,balloon,fdelete+replace+truncate
aklon,balloon,fdelete+replace+single
;

proc print data=Edit_Dist label;
   label GED='Generalized Edit Distance'
       LEV='Levenshtein Edit Distance';
   var String1 String2 GED LEV Operation;
run;
```

Output 6.4: Comparing COMPGED and COMPLEV

Obs	String1	String2	Generalized Edit Distance	Levenshtein Edit Distance	Operation
1	balloon	balloon	0	0	match
2	ba lloon	balloon	10	1	blank
3	balloo	balloon	10	1	truncate
4	baalloon	balloon	20	1	double
5	ballon	balloon	20	1	single
6	balolon	balloon	20	2	swap
7	ball.oon	balloon	30	1	punctuation
8	balloons	balloon	50	1	append
9	balkloon	balloon	100	1	insert
10	blloon	balloon	100	1	delete
11	ba1loon	balloon	100	1	replace
12	blolon	balloon	120	3	swap+delete
13	balls	balloon	120	3	replace+truncate*2
14	balXtoon	balloon	200	2	replace+insert
15	blYloon	balloon	200	2	insert+delete
16	bkakloon	balloon	200	2	insert+replace
17	bliooX	balloon	200	2	delete+replace
18	kballoon	balloon	200	1	finsert
19	alloon	balloon	200	1	fdelete
20	kalloon	balloon	200	1	freplace
21	akloon	balloon	300	2	fdelete+replace
22	akloo	balloon	310	3	fdelete+replace+truncate
23	aklon	balloon	320	3	fdelete+replace+single

As you can see from Output 6.4 above, the costs associated with GED can be quite different from costs for LEV. This is because each operation in the LEV result is equal to 1 (except for swap because it is changing two characters at once), while the computational cost associated with each operation varies for GED. So, the accumulated cost for the final result will have different scales, with more differentiation in the GED result due to the *kind* of operation, as well as the number. I will focus more on GED as it is more widely used in matching applications. But it is easy to change the function calls and testing thresholds to align with LEV, if that is the desired matching methodology.

6.3.2 SOUNDEX

Now, GED is a very useful, powerful method for comparing two strings. However, some kinds of strings, particularly names, can have improved matching rates with some additional preprocessing performed. The SOUNDEX function implements a technique for processing names that was initially developed in 1918, improved in 1922, and remains useful today.[3]

Note: It's important to recognize that SOUNDEX was developed for processing English names, and the rules inherent in the algorithm are therefore biased in favor of English names. I would recommend therefore, that you use COMPGED without SOUNDEX for non-English names.

SOUNDEX Algorithm

Below is the algorithm implemented by the SOUNDEX function to reduce match errors between names.

1. Keep the first letter of the string, and discard: A E H I O U W Y
2. Assign values to letters as follows:
   1: B F P V

2: C G J K Q S X Z

3: D T

4: L

5: M N

6: R

3. If, before any letters were discarded, there are adjacent letters of the same classification in *Step 2*, then keep only the first one.

I have created a few simple name examples using SOUNDEX below. Notice in Output 6.5 how effective the algorithm is at mapping the phonemically similar names to the same encoding. This is just a taste of how effective the algorithm is for processing English names.

```
data MyNames;
   input Name : $12.;
   SDX = Soundex(name);
   datalines;
Tom
Tommy
Tomas
Thomas
Bonny
Bony
Bonnie
Bonie
Lori
Laurie
Lauree
;

proc print data=MyNames;
label SDX="SOUNDEX Code";
run;
```

Output 6.5: Soundex Codes

Obs	Name	SDX
1	Tom	T5
2	Tommy	T5
3	Tomas	T52
4	Thomas	T52
5	Bonny	B5
6	Bony	B5
7	Bonnie	B5
8	Bonie	B5
9	Lori	L6
10	Laurie	L6
11	Lauree	L6

As you can see from Output 6.5 above, we have a greatly improved chance of getting "exact" matches for names after SOUNDEX processing. And by applying COMPGED to the resulting codes rather than the raw names, you also have improved performance. This helps when processing names that have been manually entered into the source data, as mistakes are more likely.

Caution: SOUNDEX can lead to over-matching, so use with care. Test thresholds with your data set to understand the sensitivity for your application.

6.3.3 Putting Things Together

Now it's time to show how matching different data sets actually occurs in reality. Up to this point, I've shown you toy examples of how the functions work, but not how you would then use these functions to actually take disjoint data sets and merge them together on the basis of the function results. I will cover that topic now.

Note: The examples below will not be large enough to require indexes, but I expect you can use the PII generator code provided in Appendix A to create very large data sets that will.

We can accomplish threshold-specific merges using PROC SQL in SAS code as follows. And fair warning, this kind of matching process is inherently slow, but I will take steps to make it as fast as possible. You will need to tune your processes based on the data and business context with which you are faced.

Sample PII Data

Below is the code I use to take some *randomly generated* PII data (see Appendix A), and ingest it for a processing my matching examples to follow.

```
data pii_data;
infile "C:\Users\mawind\Documents\SASBook\Examples\PII_text.csv" dsd firstobs=2
dlm=",";
input First_Name : $20. Last_Name : $20. Phone : $15. SSN : $11. DOB : mmddyy10.
Address : $100.;

*Breaking the Address field into its distinct pieces;
Street=scan(address,1,',');
City=scan(address,2,',');
State = scan(address,-2);
Zip = scan(address,-1);

*Cleaning up the phone numbers;
Clean_Phone = PRXPARSE('s/(\+|\.|\(|\)|-)//o');
CALL PRXCHANGE(Clean_Phone,-1,Phone);

*Clean up the SSNs;
Clean_SSN = PRXPARSE('s/(\+|\.|\(|\)|-)//o');
CALL PRXCHANGE(Clean_SSN,-1,SSN);

drop Clean_Phone Clean_SSN;
run;

proc print data=pii_data;
run;
```

The steps used to clean up the source files should be familiar by now, except for the SCAN function. I chose to parse my address field a different way to demonstrate some variety in approach. By using SCAN, I'm able to quickly pull out street, city, state, and zip as separate variables. You can see the regular expressions (RegEx) used later in the code to clean up the phone numbers and Social Security numbers. Below is the PROC PRINT output for the PII_DATA.

Output 6.6: PROC PRINT Output of PII_DATA

Obs	First_Name	Last_Name	Phone	SSN	DOB	Address	Street	City	State	Zip
1	ROBERT	SMITH	91921370163	449310992	9847	1776 D St NW, Washington, DC 20006	1776 D St NW	Washington	DC	20006
2	MICHAEL	JOHNSON	67781408733	181462647	7754	1600 Pennsylvania Ave NW, Washington, DC 20500	1600 Pennsylvania Ave NW	Washington	DC	20500
3	WILLIAM	WILLIAMS	32841860378	456883230	9771	600 14th St NW, Washington, DC 20005	600 14th St NW	Washington	DC	20005
4	DAVID	JONES	10139253894	666668474	9352	1321 Pennsylvania Ave NW, Washington, DC 20004	1321 Pennsylvania Ave NW	Washington	DC	20004
5	RICHARD	BROWN	39535122264	224632202	7330	2470 Rayburn Hob, Washington, DC 20515	2470 Rayburn Hob	Washington	DC	20515
6	DANIEL	DAVIS	26682053559	274704190	11828	101 Independence Ave SE, Washington, DC 20540	101 Independence Ave SE	Washington	DC	20540
7	PAUL	MILLER	31255423883	856135726	6461	511 10th St NW, Washington, DC 20004	511 10th St NW	Washington	DC	20004
8	MARK	WILSON	22891853013	775068913	7443	450 7th St NW, Washington, DC 20004	450 7th St NW	Washington	DC	20004
9	DONALD	MOORE	81590379848	916150865	11532	2 15th St NW, Washington, DC 20007	2 15th St NW	Washington	DC	20007
10	KENNETH	TAYLOR	25277623314	362986110	8067	3700 O St NW, Washington, DC 20057	3700 O St NW	Washington	DC	20057
11	BRIAN	ANDERSON	70895812036	959461229	8622	3001 Connecticut Ave NW, Washington, DC 20008	3001 Connecticut Ave NW	Washington	DC	20008
12	RONALD	THOMAS	43595665873	017219918	5986	3101 Wisconsin Ave NW, Washington, DC 20016	3101 Wisconsin Ave NW	Washington	DC	20016
13	ANTHONY	JACKSON	36177521060	825214114	6458	800 Florida Ave NE, Washington, DC 20002	800 Florida Ave NE	Washington	DC	20002
14	MARY	WHITE	55570921923	381640076	7066	1 First St NE, Washington, DC 20543	1 First St NE	Washington	DC	20543
15	PATRICIA	HARRIS	64376020129	077991188	8927	600 Independence Ave SW, Washington, DC 20560	600 Independence Ave SW	Washington	DC	20560
16	BARBARA	MARTIN	34066145946	230252012	5780	10th St. & Constitution Ave. NW, Washington, DC 20560	10th St. & Constitution Ave. NW	Washington	DC	20560
17	JENNIFER	THOMPSON	35480807899	461494171	8469	555 Pennsylvania Ave NW, Washington, DC 20001	555 Pennsylvania Ave NW	Washington	DC	20001
18	MARIA	GARCIA	01875921898	874932184	11977	1000 5th Ave, New York, NY 10028	1000 5th Ave	New York	NY	10028
19	MARGARET	MARTINEZ	00783942779	590471712	6804	64th St and 5th Ave, New York, NY 10021	64th St and 5th Ave	New York	NY	10021
20	DOROTHY	ROBINSON	76135388679	829946157	5404	10 Lincoln Center Plaza, New York, NY 10023	10 Lincoln Center Plaza	New York	NY	10023
21	LISA	CLARK	95690161356	416304984	11629	Pier 86 W 46th St and 12th Ave, New York, NY 10036	Pier 86 W 46th St and 12th Ave	New York	NY	10036
22	KAREN	RODRIGUEZ	26000627719	305140427	9493	350 5th Ave, New York, NY 10118	350 5th Ave	New York	NY	10118
23	SANDRA	LEWIS	76266560248	863529665	5509	405 Lexington Ave, New York, NY 10174	405 Lexington Ave	New York	NY	10174
24	LAURA	LEE	80119359763	884288732	9042	1 Albany St, New York, NY 10006	1 Albany St	New York	NY	10006
25	DEBORAH	WALKER	99756736434	117721843	11217	1585 Broadway, New York, NY 10036	1585 Broadway	New York	NY	10036

Next, I create PII_DATA3, a transformed subset of PII_DATA, to demonstrate how to match similar, but not exactly the same records, using just names. As you can see in the code below, I manually make edits in the first and last name fields for three records so as to demonstrate the concepts of fuzzy matching (but, not *too* fuzzy!) in the PROC SQL examples to follow.

```
data work.pii_data3;
set work.pii_data (obs=10);
if last_name='BROWN' then first_name='RICH';
if first_name='MICHAEL' then last_name='JONSON';
if first_name='PAUL' then last_name='MILER';
run;

proc print data=pii_data3;
run;
```

The output from the above PROC PRINT statement is provided in Output 6.7. As expected, I have created a data set containing the first 10 records of PII_DATA, with edits made to 3 records.

Output 6.7: PROC PRINT Output of PII_DATA3

Obs	First_Name	Last_Name	Phone	SSN	DOB	Address	Street	City	State	Zip
1	ROBERT	SMITH	91921370163	449310992	9847	1776 D St NW, Washington, DC 20006	1776 D St NW	Washington	DC	20006
2	MICHAEL	JONSON	67781408733	181462647	7754	1600 Pennsylvania Ave NW, Washington, DC 20500	1600 Pennsylvania Ave NW	Washington	DC	20500
3	WILLIAM	WILLIAMS	32841860378	456883230	9771	600 14th St NW, Washington, DC 20005	600 14th St NW	Washington	DC	20005
4	DAVID	JONES	10139253894	666668474	9352	1321 Pennsylvania Ave NW, Washington, DC 20004	1321 Pennsylvania Ave NW	Washington	DC	20004
5	RICH	BROWN	39535122264	224632202	7330	2470 Rayburn Hob, Washington, DC 20515	2470 Rayburn Hob	Washington	DC	20515
6	DANIEL	DAVIS	26682063559	274704190	11828	101 Independence Ave SE, Washington, DC 20540	101 Independence Ave SE	Washington	DC	20540
7	PAUL	MILER	31255423883	856135726	6461	511 10th St NW, Washington, DC 20004	511 10th St NW	Washington	DC	20004
8	MARK	WILSON	22891853013	775068913	7443	450 7th St NW, Washington, DC 20004	450 7th St NW	Washington	DC	20004
9	DONALD	MOORE	81590379848	916150865	11532	2 15th St NW, Washington, DC 20007	2 15th St NW	Washington	DC	20007
10	KENNETH	TAYLOR	25277623314	362986110	8067	3700 O St NW, Washington, DC 20057	3700 O St NW	Washington	DC	20057

Now, to create a baseline for comparison in the following examples, I have provided a PROC SQL statement below that generates only output (i.e., doesn't store the results in a table). This is to help you see the exact matching case for first and last names in our sample data sets.

You see that I'm selecting only the first name and last name from each data set (renaming the first and last name columns from PII_DATA3 to avoid an error). And I'm creating the exact matching conditions in my WHERE clause with an AND statement to ensure that both equality conditions are met. Also, notice the use of the QUIT statement in lieu of the RUN statement. This is required to exit PROC SQL.

```
proc sql;
select A.first_name, A.last_name,
B.first_name as Fname, B.last_name as Lname

from work.pii_data A, work.pii_data3 B
where A.last_name = B.last_name
AND A.first_name = B.first_name;
quit;
```

Note: After the PROC SQL statement is submitted, you can run as many SQL statements as you want—RUN has no effect—until you submit the QUIT statement to end SQL processing.

Output 6.8: Exact Match Results

First_Name	Last_Name	Fname	Lname
ROBERT	SMITH	ROBERT	SMITH
WILLIAM	WILLIAMS	WILLIAM	WILLIAMS
DAVID	JONES	DAVID	JONES
DANIEL	DAVIS	DANIEL	DAVIS
MARK	WILSON	MARK	WILSON
DONALD	MOORE	DONALD	MOORE
KENNETH	TAYLOR	KENNETH	TAYLOR

As you can see in Output 6.8, we get only 7 exact matches, which was expected given that I made edits to 3 records. Note that those edited records do not appear in the output.

Now that I have established a baseline with an exact match example, I'm going to incorporate COMPGED into the WHERE clause in lieu of the equal statements.

```
proc sql;
select A.first_name, A.last_name,
B.first_name as Fname, B.last_name as Lname
```

```
from work.pii_data A, work.pii_data3 B
where compged(A.first_name,B.first_name)<50
AND compged(A.last_name,B.last_name)<50;
quit;
```

You can see that I placed the COMPGED function in the WHERE clause above, comparing the first and last name for each data set. I established an arbitrary cost of 50 for my threshold (See Section 6.3.1 for the COMPGED cost list). This allows for a fair amount of fuzziness. And it results in an additional match ("Paul Miller" and "Paul Miler").

Output 6.9: COMPGED Match with 50 Threshold

First_Name	Last_Name	Fname	Lname
ROBERT	SMITH	ROBERT	SMITH
WILLIAM	WILLIAMS	WILLIAM	WILLIAMS
DAVID	JONES	DAVID	JONES
DANIEL	DAVIS	DANIEL	DAVIS
PAUL	MILLER	PAUL	MILER
MARK	WILSON	MARK	WILSON
DONALD	MOORE	DONALD	MOORE
KENNETH	TAYLOR	KENNETH	TAYLOR

I can continue adjusting the thresholds up until I get the additional records, but I would do so at the risk of over-matching—a big problem in the real world. So, I can use SOUNDEX to improve the power of my name-matching approach without too much additional risk (again, this is used for English names rather than general applications).

After adding SOUNDEX inside the COMPGED function calls, I am able to convert the first and last name variables on the fly for the comparison. I recommend running SOUNDEX to actually create new variables for large-scale applications, but this works without noticeable performance impact for small (less than three memory pages) data sets. Notice that my new thresholds are now 20 for both the first and last name comparisons. So, my COMPGED cost associated with SOUNDEX converted variables must be less than 20, which is quite reasonable.

```
proc sql;
select A.first_name, A.last_name,
B.first_name as Fname, B.last_name as Lname

from work.pii_data A, work.pii_data3 B
where compged(soundex(A.first_name),soundex(B.first_name))<20
AND compged(soundex(A.last_name),soundex(B.last_name))<20;
quit;
```

These updates result in Output 6.10 below, which shows that we get 9 matches for lower COMPGED cost values of 20 for both comparisons ("Michael Johnson" and "Michael Jonson").

Output 6.10: Combining SOUNDEX and COMPGED

First_Name	Last_Name	Fname	Lname
ROBERT	SMITH	ROBERT	SMITH
MICHAEL	JOHNSON	MICHAEL	JONSON
WILLIAM	WILLIAMS	WILLIAM	WILLIAMS
DAVID	JONES	DAVID	JONES
DANIEL	DAVIS	DANIEL	DAVIS
PAUL	MILLER	PAUL	MILER
MARK	WILSON	MARK	WILSON
DONALD	MOORE	DONALD	MOORE
KENNETH	TAYLOR	KENNETH	TAYLOR

While this does not remove risk of over-matching, it does better control for the kind of spelling errors that we may naturally see in hand-coded name values in real-world data sources. So, this enables you to account for many of those potential issues in English names, and then narrow the COMPGED tolerance to a much smaller level. This ensures you don't accept too many uncommon errors in name fields in pursuit of fuzzy matching.

Notice that I still can't get Richard to show up in the resulting data set above (see Output 6.10). The truncation down to "Rich" is large enough that I would need to increase the threshold for the COMPGED results. Now, I don't have to increase the threshold for both first and last name comparisons.

It is more common for nicknames to creep into a customer relationship management (CRM) system, for example. So, below are the results of increasing the threshold for first name to 150, while keeping the last name threshold at 20. You can see that I am able to match all ten names in PII_DATA3 after just making the match criteria for First_Name much looser.

Output 6.11: Full Match with SOUNDEX and COMPGED

First_Name	Last_Name	Fname	Lname
ROBERT	SMITH	ROBERT	SMITH
MICHAEL	JOHNSON	MICHAEL	JONSON
WILLIAM	WILLIAMS	WILLIAM	WILLIAMS
DAVID	JONES	DAVID	JONES
RICHARD	BROWN	RICH	BROWN
DANIEL	DAVIS	DANIEL	DAVIS
PAUL	MILLER	PAUL	MILER
MARK	WILSON	MARK	WILSON
DONALD	MOORE	DONALD	MOORE
KENNETH	TAYLOR	KENNETH	TAYLOR

I believe the above steps demonstrate how you can potentially tune a matching approach to account for a wide variety of desired outcomes. Now, to take this discussion to a close, I think it is necessary to demonstrate how you might want to construct a stronger set of matches. Clearly, name matches alone would not be sufficient for resolving entity references in many real-world applications. So, I'm going to use some of the other variables to truly resolve the entity references.

Combining Full Name, DOB, and SSN is generally accepted as a means of creating a strong match between entity references. I just walked through how you can tackle the name matching portion of the exercise, but the other elements are not without their challenges.

The SSN is a system-generated element passed across enterprise data systems with relative ease, and little or no human interaction. However, there are numerous times with people enter their SSN into an online or paper form. That creates the possibility of transposing digits, or other errors. As a result, you may occasionally want to provide for a *limited* window of error, depending on your application; but I would not recommend it as a general rule. If you were to, for whatever reason, loosen the strictness of matching for the SSN variable, it is important to ensure it is being stored as a text field. You can't perform edit distance on a numeric field, and the numerical distance between two SSNs doesn't make sense here. Simply switching two numbers could create numbers that are very far apart, numerically for a minor "fat finger" error.

DOB on the other hand, while having many of the same characteristics, isn't meant to uniquely identify a person in a government system of record. So, it must be tied to a name or other PII element to be useful as a matching constraint. Also, as long as the DOB is stored as a date number, it makes sense to perform distance calculations for the purposes of fuzzy matching with it. However, if the source fields were character fields, it is possible for edit distance to be more relevant. Understanding your data lineage is important for identifying potential issues, and developing strategies for resolving them.

For our purposes here, I will wrap up by completing my matching SQL with the generally preferred method for matching these elements: exact match on both SSN and DOB. That, combined with mild fuzzy matches on names, generates a reasonable data set for us to use.

Note: Your particular context may not allow even this level of fuzzy matching for any elements of PII data, but you can still use these concepts for performing entity network mapping and analysis in the coming chapters.

In the updated code below, I have added the CREATE TABLE statement in order to generate a SAS data set (WORK.MERGE_FUZZY) from my PROC SQL. I have also added fields to the SELECT statement so that they can be displayed or output to the resulting data set. I'm also adding the INNER JOIN statement here as it is a best practice for joining tables and is more efficient than using a simple WHERE clause.

```
proc sql;
create table work.merge_fuzzy as

select A.first_name, A.last_name, A.SSN, A.DOB,
B.first_name as Fname, B.last_name as Lname, B.SSN as Social, B.DOB as BirthDate

from work.pii_data A inner join work.pii_data3 B
    on compged(soundex(A.first_name),soundex(B.first_name))<150
    AND compged(soundex(A.last_name),soundex(B.last_name))<20
    AND A.DOB = B.DOB
    AND A.SSN = B.SSN
;
quit;

proc print data=work.merge_fuzzy;
run;
```

Just to explain the above syntax, the FROM clause is essentially saying: Perform an INNER JOIN A and B, ON all of the following criteria.

Novice Note: The syntax of SQL is a little confusing at first, but its efficiency for running large-scale data manipulation makes it worth the additional effort. Since you can do it in a familiar environment like SAS, I would highly recommend taking the time to learn it.

As you can see in the output below, I still get a match on all 10 records in common between the data sets. You can see the unformatted DOB date number and unformatted SSN text in each record, along with the same names that we have been experimenting with in this section.

Output 6.12: PROC PRINT Output of MERGE_FUZZY

Obs	First_Name	Last_Name	SSN	DOB	Fname	Lname	Social	BirthDate
1	ROBERT	SMITH	449310992	9847	ROBERT	SMITH	449310992	9847
2	MICHAEL	JOHNSON	181462647	7754	MICHAEL	JONSON	181462647	7754
3	WILLIAM	WILLIAMS	456883230	9771	WILLIAM	WILLIAMS	456883230	9771
4	DAVID	JONES	666668474	9352	DAVID	JONES	666668474	9352
5	RICHARD	BROWN	224632202	7330	RICH	BROWN	224632202	7330
6	DANIEL	DAVIS	274704190	11828	DANIEL	DAVIS	274704190	11828
7	PAUL	MILLER	856135726	6461	PAUL	MILER	856135726	6461
8	MARK	WILSON	775068913	7443	MARK	WILSON	775068913	7443
9	DONALD	MOORE	916150865	11532	DONALD	MOORE	916150865	11532
10	KENNETH	TAYLOR	362986110	8067	KENNETH	TAYLOR	362986110	8067

Now, if I had set an overall match threshold at the beginning of this exercise of 100%, I would have just met that threshold as I have achieved the matching criteria for each individual variable in the data sets. But, this was possible only because I increased the level of fuzziness for the matches between first and last names. If I still wanted to maintain a 100% overall match threshold without including fuzzy matching on names, I would have obtained a data set with only 7 records.

Determining whether that level of fuzzy match is acceptable for names, or any variable, is driven entirely by the context in which you are working. As I said at the beginning of this book, a plan for how to handle these decisions should be developed at the outset of any project. Experiments like those we have done here may be necessary to answer initial questions by management during the planning phase, which is reasonable to do with representative sample data.

Surviving References

Output 6.12 shows the two entity references matched together in my final data set. But one question is left unresolved: Which version of the entity reference do I use? The answer is: Well, it depends.

Generally speaking, you will have a system of record, or other trusted system against which you are testing potential matches. For example, you might have a database of donors and a table of recent donations; you would want to match the recent donations against the clean database of donor information. In that kind of situation, you would keep the donor database system of record, while adding the donation to their donation records.

However, there could be situations in which you are pulling together two sources of data that are external to your organization, with no point of reference. Without a baseline, you will have to research your sources, and determine the source that you trust the most. Document that information for the person who has requested your analysis, and use the version of an entity reference within that source as the *surviving reference*. This does not always lead to satisfying answers, especially when building up a reference database from scratch. But you can, with enough storage space, keep the ambiguous references for reprocessing as you obtain additional data. When multiple sources show an entity reference combination that agrees, you can accept that version as the correct version for future use.

6.4 Summary

I finally got to the entity resolution part of this book! Now you know how to go from potentially very raw data sources to actually creating entity reference matches. However, as I mentioned earlier, that is not where many real-world applications end. As shown in the next chapter, Entity Network Mapping and Analysis, the matching process can be applied to create unresolved links that we want to preserve for a variety of applications. But why?! You'll have to read the next chapter to see.

---

[1]SAS Institute Inc. "Understanding SAS Indexes," *SAS® 9.2 Language Reference: Concepts, Second Edition,*
http://support.sas.com/documentation/cdl/en/lrcon/62955/HTML/default/viewer.htm#a000440261.htm (accessed August 29, 2018).

[2] SAS Institute Inc. "COMPGED Function," *SAS® 9.4 Functions and CALL Routines: Reference, Fifth Edition,*
http://go.documentation.sas.com/?docsetId=lefunctionsref&docsetTarget=p1r4l9jwgatggtn1ko81fyjys4s7.htm&docsetVersion=9.4&locale=en (accessed August 29, 2018).

[3] SAS Institute Inc. "SOUNDEX Function," *SAS® 9.4 Functions and CALL Routines: Reference, Fifth Edition,*
http://go.documentation.sas.com/?docsetId=lefunctionsref&docsetTarget=n1i9a3o4kciemhn1kpgutl20e4i0.htm&docsetVersion=9.4&locale=en (accessed August 29, 2018).

Chapter 7: Entity Network Mapping and Analysis

7.1 Introduction

This chapter will focus on structuring and analyzing your data to understand the networks formed by entity relationships within your data sources. When performing Entity Resolution (ER), as I did in the last chapter, I am attempting to resolve entity references to the same real-world entity. However, Entity Network Mapping and Analysis (ENMA) is designed to establish and characterize linkages between different real-world entities based on shared characteristics of the entity references. In other words, I don't want to merge the entity references into a single record or reference. Instead, I just want to denote a connection between them using entity reference attributes. It is important to do this after entity resolution, as I don't want to create links between entities that ultimately get fused together as one.

Figure 7.1: ERA Flow with Entity Network Mapping and Analysis Focus

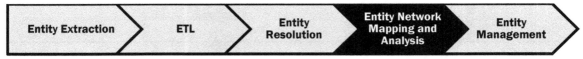

All discussion in the last chapter ignored the issue of entity type (i.e., person, corporation, etc.) as I was focused on just *resolving* entity references. In real-world applications, heterogeneous sets of entity references are often dealt with as part of the broader ERA project. In other words, the *end goal* of an ERA project isn't only to see whether entities of the same type are in fact the same person, place, or thing; instead, it is to understand the relationship among many distinctly different entities. And context plays a major role in the decision to comingle entity types for this purpose.

For instance, it is more common—even necessary—in the national security and law enforcement regimes to comingle entity reference types for the benefit of investigations, but it is less common in other regimes.

Example: An FBI special agent needs to create a holistic picture of a suspect's life, including all the entities with which he or she is associated: family members, vehicles, businesses, coworkers, credit cards, bank accounts, and so on. Therefore, comingling different entity types clearly serves a business need.

Now, links identified between entity references are again driven by context, and the associated business decisions. This linkage information should be recast as a multi-dimensional, heterogeneous network graph to best support any future analysis. In other words, we can assume that a network graph is formed with any number of entity types (comingling). As such, all of the entities become "nodes," while the attributes that link them become the "edges" of the graph. The graph information is stored separately from the ER results so as to avoid improper integration of our resolved entities and linked entities. This graph can then be leveraged for visualization and analysis.

Linkages between multiple entities form networks, which are important for a variety of applications, including fraud and criminal investigations. This information can also be extremely informative for very different purposes, such as social media ad campaigns.

Rigorous analysis of entity networks can lead to discoveries that are easily obscured by the large volumes of data present--discoveries such as fraud rings, identity theft, and social media market penetration. This analysis must be automated because the data volumes tend to be so large as to make manual analysis infeasible. And the analytical algorithms applied at this stage are dependent upon the specific business need.

There are some very sophisticated procedures for analyzing networks within the SAS platform; however, the more sophisticated options are bundled into enterprise solutions such as SAS Fraud Framework. Since I want to introduce you to the most widely available tools possible, I have focused on procedures that are available in SAS/OR, a more generally available software suite that is more likely to be licensed by either you or your employer.

7.2 Entity Network Mapping

The entity network mapping phase of the ERA process is, as usual, driven by the specific business needs and available data. There are essentially two ways for networks of entities to be identified in your data sources:

- Shared entity attributes
- Entity interactions

7.2.1 Shared Entity Attributes

When two entity references share one or more attributes, but not enough to be resolved as the same entity reference, then you are able only to identify a relationship between them. The number and type of attributes they share will shape how you characterize that relationship. The analyses demonstrated in Section 7.3 will provide additional opportunities to characterize these relationships, and the networks formed by them.

Now, the shared entity attributes discussion is the most straightforward as it is a very similar process to that of entity resolution, but with a different goal in mind.

There are many examples of entities that share attributes, such as a husband and wife sharing an address or phone number, while not actually being the same entity. And I want to identify as many vectors along which to establish and characterize such relationships in the soup of entity data I have available. However, I still have to apply some logical boundaries in which to operate. That means that I'm not going to check every entity reference attribute against every other, as that would be counterproductive. Rather, I want to check reference attributes that could logically establish a relationship in geographical, temporal, or group identity vectors. There are no hard and fast rules for how to choose these, but knowledge of your application can guide logical research and decision-making here.

For example, I obviously don't want to establish links between all individuals named Bob, as that wouldn't be useful information, especially since I've already resolved my entity references. Instead, I would want to use attributes like address, phone number, or employer that allow me to establish and characterize a relationship among entities. A collection of such links will form entity networks.

Below is a reminder of the PII data set I used quite a bit in Chapter 6, and the process for mapping shared entity attributes is the same as resolving entities like I did there; however, I would do it only one attribute at a time. Given the similarity, I don't feel the need to repeat that work here as you have the tools to do it already. However, I do want to make a few notes regarding the potential for network development.

Figure 7.1: PII Data Set Reminder

Obs	First_Name	Last_Name	Phone	SSN	DOB	Address	Street	City	State	Zip
1	ROBERT	SMITH	91921370163	449310992	9847	1776 D St NW, Washington, DC 20006	1776 D St NW	Washington	DC	20006
2	MICHAEL	JOHNSON	67761408733	181462647	7754	1600 Pennsylvania Ave NW, Washington, DC 20500	1600 Pennsylvania Ave NW	Washington	DC	20500
3	WILLIAM	WILLIAMS	32841860378	456883230	9771	600 14th St NW, Washington, DC 20005	600 14th St NW	Washington	DC	20005
4	DAVID	JONES	10139253894	666666474	9352	1321 Pennsylvania Ave NW, Washington, DC 20004	1321 Pennsylvania Ave NW	Washington	DC	20004
5	RICHARD	BROWN	39535122264	224632202	7330	2470 Rayburn Hob, Washington, DC 20515	2470 Rayburn Hob	Washington	DC	20515
6	DANIEL	DAVIS	26682053559	274704190	11828	101 Independence Ave SE, Washington, DC 20540	101 Independence Ave SE	Washington	DC	20540
7	PAUL	MILLER	31255423883	856135726	6461	511 10th St NW, Washington, DC 20004	511 10th St NW	Washington	DC	20004
8	MARK	WILSON	22891853013	775068913	7443	450 7th St NW, Washington, DC 20004	450 7th St NW	Washington	DC	20004
9	DONALD	MOORE	81590379646	916150865	11532	2 15th St NW, Washington, DC 20007	2 15th St NW	Washington	DC	20007
10	KENNETH	TAYLOR	25277623314	362986110	8067	3700 O St NW, Washington, DC 20057	3700 O St NW	Washington	DC	20057
11	BRIAN	ANDERSON	70895812036	959461229	8622	3001 Connecticut Ave NW, Washington, DC 20008	3001 Connecticut Ave NW	Washington	DC	20008
12	RONALD	THOMAS	43595655873	017219918	5986	3101 Wisconsin Ave NW, Washington, DC 20016	3101 Wisconsin Ave NW	Washington	DC	20016
13	ANTHONY	JACKSON	36177521060	825214114	6458	800 Florida Ave NE, Washington, DC 20002	800 Florida Ave NE	Washington	DC	20002
14	MARY	WHITE	55570921923	381640076	7066	1 First St NE, Washington, DC 20543	1 First St NE	Washington	DC	20543
15	PATRICIA	HARRIS	64375020129	077991188	8927	600 Independence Ave SW, Washington, DC 20560	600 Independence Ave SW	Washington	DC	20560
16	BARBARA	MARTIN	34066145946	230252012	5780	10th St. & Constitution Ave NW, Washington, DC 20560	10th St. & Constitution Ave NW	Washington	DC	20560
17	JENNIFER	THOMPSON	35480807899	461494171	8469	555 Pennsylvania Ave NW, Washington, DC 20001	555 Pennsylvania Ave NW	Washington	DC	20001
18	MARIA	GARCIA	01875921898	874932184	11977	1000 5th Ave, New York, NY 10028	1000 5th Ave	New York	NY	10028
19	MARGARET	MARTINEZ	00783942779	590471712	6804	64th St and 5th Ave, New York, NY 10021	64th St and 5th Ave	New York	NY	10021
20	DOROTHY	ROBINSON	76135389679	829946157	5404	10 Lincoln Center Plaza, New York, NY 10023	10 Lincoln Center Plaza	New York	NY	10023
21	LISA	CLARK	95690161356	416304964	11629	Pier 86 W 46th St and 12th Ave, New York, NY 10036	Pier 86 W 46th St and 12th Ave	New York	NY	10036
22	KAREN	RODRIGUEZ	26000627719	305140427	9493	350 5th Ave, New York, NY 10118	350 5th Ave	New York	NY	10118
23	SANDRA	LEWIS	76266560248	863529665	6509	405 Lexington Ave, New York, NY 10174	405 Lexington Ave	New York	NY	10174
24	LAURA	LEE	80119359763	884286732	9042	1 Albany St, New York, NY 10006	1 Albany St	New York	NY	10006
25	DEBORAH	WALKER	99756736434	117721843	11217	1585 Broadway, New York, NY 10036	1585 Broadway	New York	NY	10036

The entity networks formed by this activity can easily be heterogeneous in nature. As you might imagine, entity references of very different types can share attributes of interest to investigators or researchers.

For example, you may have a database table of cell phones, as well as a table of people. Assuming that both of these data tables have phone number as an attribute, you could use that share reference attribute about the cell phones and the persons to generate a network of cell phones and persons. This is a very simplistic example of how you can blend entity types as a means of revealing indirect relationships between real-world entities of interest (usually persons). Depending on the context of application, the construction of these heterogeneous networks can reveal anything from fraud rings to cancer clusters.

7.2.2 Entity Interactions

In addition to sharing attributes, such as address or phone number, entity references will appear that characterize transactions where real-world entities are interacting based on the data (for example, a call log showing just the phone numbers of a caller and recipient). This kind of data creates a record of interaction between the phone number owners. So, I can use such information to establish a link between real-world entities based on this kind of interaction between entity reference attributes.

In addition to simply having attributes in common, there are many instances when systems contain information that demonstrates interactions between different entities. The types of interactions, and data sources describing them, will determine what you will want to do with said information—all within the business context. Since this becomes very context specific, and I can't hope to cover every context, I will proceed by way of example. I believe you can extrapolate effectively from the following.

In order to effectively demonstrate the interaction between entities, I have manually created a call log using the phone numbers in the PII data set shown in Figure 7.2. Below is a sample of the 200 calls I manually generated via copy and paste in a spreadsheet.

Figure 7.2: Phone Call Log Sample

Obs	From	To
1	91921370163	36177521060
2	67781408733	55570921923
3	32841860378	64375020129
4	10139253894	34066145946
5	39535122264	35480807899
6	26682053559	01875921898
7	31255423883	00783942779
8	22891853013	76135388679
9	81590379848	95690161356
10	25277623314	26000627719
11	70895812036	76266560248
12	43595655873	80119359763

Now, analyzing the call log information without the associated entity references is interesting, but less informative, for my end goal of understanding how the entities are related. So, in the code below, I am performing a left join with the originating data set pictured in Figure 7.2.

```
proc sql;
create table work.temp as
select B.From, B.To, A.last_name as ToName
from work.pii_data A left join work.phone_calls B
    on A.phone = B.To
;

create table work.CallMap as
select B.last_name as FromName, A.*
from work.temp A left join work.pii_data B
    on A.From = B.phone
    order by FromName
;
quit;

proc print data=work.CallMap;
run;
```

Output 7.1: Join of Names and Calls

Obs	FromName	From	To	ToName
1	ANDERSON	70895812036	80119359763	LEE
2	ANDERSON	70895812036	80119359763	LEE
3	ANDERSON	70895812036	80119359763	LEE
4	ANDERSON	70895812036	76266560248	LEWIS
5	ANDERSON	70895812036	76266560248	LEWIS
6	ANDERSON	70895812036	80119359763	LEE
7	ANDERSON	70895812036	80119359763	LEE
8	BROWN	39535122264	01875921898	GARCIA
9	BROWN	39535122264	01875921898	GARCIA
10	BROWN	39535122264	35480807899	THOMPSON
11	BROWN	39535122264	01875921898	GARCIA
12	BROWN	39535122264	35480807899	THOMPSON

Now that I have created a mapping between known entities, I want to characterize that mapping in advance of doing any network analysis. The first characterization of the network will be the count of connections among callers using PROC FREQ. Note that the PROC FREQ output is a matrix with call counts between caller and recipient. I used the options NOPERCENT, NOCUM, NOROW, and NOCOL to generate this simplified view. Also, while this is the view in my output below, it is a simplified structure for the data set being stored as WORK.CALLMAP rather than a matrix.

```
proc freq data=work.callmap;
tables fromname*toname / out=work.counts nopercent nocum nocol norow;
run;
```

Output 7.2: Sample PROC FREQ Phone Log Counts

FromName	ANDERSON	BROWN	CLARK	DAVIS	GARCIA	HARRIS	JACKSON	JOHNSON	JONES	LEE	LEWIS	MARTIN	MARTINEZ	MILLER	MOORE	ROBINSON	RODRIGUEZ	SMITH	TAYLOR
ANDERSON	0	0	0	0	0	0	0	0	0	5	2	0	0	0	0	0	0	0	0
BROWN	0	0	0	0	5	0	0	0	0	0	0	0	0	0	0	0	0	0	0
CLARK	0	0	0	0	0	5	0	0	0	0	0	0	0	6	0	0	0	0	0
DAVIS	0	0	0	0	2	0	0	0	0	0	0	0	5	0	0	0	0	0	0
GARCIA	0	0	0	3	0	0	0	0	0	0	0	0	0	0	0	0	0	0	0
HARRIS	0	0	0	0	0	0	0	0	0	0	0	0	0	0	4	0	0	0	0
JACKSON	0	0	0	0	0	0	0	0	0	0	0	0	0	4	0	0	0	3	0
JOHNSON	0	0	0	0	0	4	0	0	0	0	0	0	0	0	0	0	0	0	0
JONES	0	0	0	0	0	4	0	0	0	0	0	2	0	0	0	0	0	0	0
LEE	0	0	0	0	5	0	0	0	0	0	0	0	0	0	0	0	0	0	0
LEWIS	6	0	0	0	0	0	0	0	0	0	0	0	0	0	0	0	0	0	0
MARTIN	0	0	0	0	0	0	0	0	3	0	0	0	0	0	0	0	0	0	5
MARTINEZ	0	0	0	0	0	0	5	0	0	0	0	0	0	3	0	0	0	0	0
MILLER	0	0	0	0	0	0	0	0	0	0	0	0	2	0	5	0	0	0	0
MOORE	0	0	2	0	0	0	0	0	0	0	0	0	0	0	0	0	5	0	0
ROBINSON	0	0	0	0	0	0	0	0	0	0	0	0	0	0	0	0	0	0	0
RODRIGUEZ	0	0	0	0	0	0	0	0	0	0	5	0	0	0	0	0	0	0	6
SMITH	0	0	0	0	0	0	2	0	0	0	0	0	0	0	0	0	0	0	0
TAYLOR	0	0	0	0	0	0	0	0	0	5	0	0	0	0	0	0	2	0	0
THOMAS	0	0	0	0	0	0	0	0	2	0	0	0	0	0	0	0	2	0	0
THOMPSON	5	3	0	0	0	0	0	0	0	0	0	0	0	0	0	0	0	0	0
WALKER	0	0	0	0	0	4	5	0	0	0	0	0	0	0	0	0	0	0	0

For the particular example I'm working with here, these counts will be used as network link weights because that is just one way I can quantify the connection between these entities. However, there are different applications and analyses that may have weights based on very different information, such as the cost of traversing the network linkages or the distance these individuals live apart.

As another example, suppose a prior analysis has determined that the count of calls between individuals has a nonlinear impact on the quality of their relationship. So, a count 6 would be far more interesting than a count 2, for example. Under that scenario, we would want to apply a function to the counts in order to create values in a matrix similar to Output 7.2, which would have incorporated this weighting in our matrix.

Again, context and business needs and available data drive how you will approach this phase of work; but whatever you do will be a corollary to what I do herein. Just remember that the entire point of this step is to attempt to quantify the relationship between entities in a meaningful way. You are wading into an area of analytics with a lot of gray. There will rarely be a clean answer to questions regarding the absolute best way to quantify relationships for your particular application—it will require interaction with subject matter experts in your domain and experience.

7.3 Entity Network Analysis

There are many different analyses that you can execute on networks formed by the shared attributes or interaction records of entities, from the most basic to the incredibly sophisticated. I will demonstrate just a few network analysis techniques available to you in the SAS/OR suite, and discuss a few additional analytical techniques that you may want to investigate further.

As with anything through this process, you will have to think very carefully about the business goals of any ERA effort, and the data that you have to support the analysis work to be performed. For these reasons, I am doing only a sampling of popular analyses, and hope you find them useful for your particular problem.

7.3.1 Articulation Points and Biconnected Components

There are a number of applications, such as managing power grids, telecom networks, or supply chains, where it can be very important for you to know the articulation points. That will be clear as I go through a national security example below.

Articulation points: Nodes in an undirected network graph that, once removed, disconnect the network, creating two or more disconnected subnetworks. In other words, these are points of failure in your network.

The below example borrows heavily from an example used in the SAS documentation for PROC OPTNET as it is a great way to demonstrate the biconnected graph and articulation point analysis available with the procedure (the data is very interesting).

After the terrorist attacks on September 11, 2001, much research and analysis was performed on the individuals involved, their contacts, and the broader community of supporters. Many publications have published the entire list, and resulting network analysis. In the DATALINES below is the sample published in the documentation for SAS/OR 14.3.[1] This provides sufficient information to demonstrate the concept of articulation point analysis in an entity network.

```
data work.TerrorNetwork;
   input from & $32. to & $32.;
   datalines;
Abu Zubeida              Djamal Beghal
Jean-Marc Grandvisir     Djamal Beghal
Nizar Trabelsi           Djamal Beghal
Abu Walid                Djamal Beghal
Abu Qatada               Djamal Beghal
Zacarias Moussaoui       Djamal Beghal
Jerome Courtaillier      Djamal Beghal
Kamel Daoudi             Djamal Beghal
Abu Walid                Kamel Daoudi
```

```
Abu Walid                    Abu Qatada
Kamel Daoudi                 Zacarias Moussaoui
Kamel Daoudi                 Jerome Courtaillier
Jerome Courtaillier          Zacarias Moussaoui
Jerome Courtaillier          David Courtaillier
Zacarias Moussaoui           David Courtaillier
Zacarias Moussaoui           Ahmed Ressam
Zacarias Moussaoui           Abu Qatada
Zacarias Moussaoui           Ramzi Bin al-Shibh
Zacarias Moussaoui           Mahamed Atta
Ahmed Ressam                 Haydar Abu Doha
Mehdi Khammoun               Haydar Abu Doha
Essid Sami Ben Khemais       Haydar Abu Doha
Mehdi Khammoun               Essid Sami Ben Khemais
Mehdi Khammoun               Mohamed Bensakhria
;
```

I created the basic data set of connections, TerrorNetwork, in the DATA step above. And with that data set, I can use PROC OPTNET below to analyze the articulation points. Using the DATA_LINKS option, I define the data set of links to be TerrorNetwork, and I create the data set NodeOut using the OUT_NODES option. This data set contains the nodes and their articulation status artpoint. The articulation points are found with PROC OPTNET by invoking the BICONCOMP statement, as I have done below. Note that this statement works only on undirected network graphs. This is intuitive when you understand that this option is just looking at the connection between network nodes, regardless of the directionality of that relationship.

```
proc optnet
    Data_Links = work.TerrorNetwork
    Out_Links = work.LinkOut
    Out_Nodes = work.NodeOut;
    biconComp;
run;
```

Now, in addition to generating the articulation points of my network, BICONCOMP also provides the biconnected components of an undirected network. You can see in the above code that I am using the OUT_LINKS option, generating the data set LinkOut, which will contain the mapping of my biconnected components in the network with the variable biconcomp. The biconnected components are numbered in order of discovery, and each link (from/to relationship) in my data set is then assigned to the affiliated component.

Biconnected components: Connected subnetworks that cannot be broken into disconnected parts by deleting any single node.

So, while articulation points reveal points of brittleness in your network, revealing biconnected components helps you understand your robust subnetworks. Depending on your analytical goals, you may need to identify these subnetworks separately to understand the ideal path for new linkages to increase redundancy for a more robust overall network. However, in the terror network case that we are dealing with here, this helps you identify closely tied subgroups or "cells" within the network. As terrorist group tactics have evolved over time (they change rapidly in response to law enforcement efforts to thwart them), this concept may become less dependable.

The articulation points identified in my network by PROC OPTNET are flagged in the NodeOut data set with the binary variable Artpoint. So, I can view only my articulation points with a PROC PRINT statement and a WHERE clause as shown in my code below. The result is shown in Output 7.3 below.

```
proc print data=work.NodeOut;
    where artpoint=1;
run;
```

Output 7.3: Articulation Points

Obs	node	artpoint
2	Djamal Beghal	1
7	Zacarias Moussaoui	1
11	Ahmed Ressam	1
14	Haydar Abu Doha	1
15	Mehdi Khammoun	1

PROC PRINT creates the results that you see below in Output 7.4. Notice that this output does not need filtering as it is just printing each link in the original data set along with the component membership in the biconcomp column. In the context of our working example, this component membership mapping effectively shows the subgroups of our network.

```
proc print data=work.LinkOut;
run;
```

Output 7.4: Biconnected Components in Terror Network Example

Obs	from	to	biconcomp
1	Abu Zubeida	Djamal Beghal	10
2	Jean-Marc Grandvisir	Djamal Beghal	1
3	Nizar Trabelsi	Djamal Beghal	2
4	Abu Walid	Djamal Beghal	9
5	Abu Qatada	Djamal Beghal	9
6	Zacarias Moussaoui	Djamal Beghal	9
7	Jerome Courtaillier	Djamal Beghal	9
8	Kamel Daoudi	Djamal Beghal	9
9	Abu Walid	Kamel Daoudi	9
10	Abu Walid	Abu Qatada	9
11	Kamel Daoudi	Zacarias Moussaoui	9
12	Kamel Daoudi	Jerome Courtaillier	9
13	Jerome Courtaillier	Zacarias Moussaoui	9
14	Jerome Courtaillier	David Courtaillier	9
15	Zacarias Moussaoui	David Courtaillier	9
16	Zacarias Moussaoui	Ahmed Ressam	6
17	Zacarias Moussaoui	Abu Qatada	9
18	Zacarias Moussaoui	Ramzi Bin al-Shibh	7
19	Zacarias Moussaoui	Mahamed Atta	8
20	Ahmed Ressam	Haydar Abu Doha	5
21	Mehdi Khammoun	Haydar Abu Doha	4
22	Essid Sami Ben Khemais	Haydar Abu Doha	4
23	Mehdi Khammoun	Essid Sami Ben Khemais	4
24	Mehdi Khammoun	Mohamed Bensakhria	3

The preceding outputs provide incredibly valuable information when making decisions about real-world networks. Whether you are attempting to shut down terror network communications or improve the robustness of your supply chain, the above analysis will give you the basic information that you need to make sound, fact-based decisions.

7.3.2 Minimum Spanning Trees

Imagine you have been tasked with designing a system where each node in the network can reach every other node in the network, using existing infrastructure, while minimizing the cost to do so. This is the classic problem solved with minimum spanning trees, and running PROC OPTNET to solve for it is relatively straightforward so long as you have the proper data. In other words, you need to have data that relates to the cost of traversing or connecting that network as the minimum spanning tree algorithm is attempting to minimize something.

Minimum spanning tree: a subset of links in a network that connects all the nodes together without any cycles, and with minimum total cost.

Below is a simple example of the classic use for minimum spanning tree analysis. I have constructed by hand a simple data set that you can follow to help you understand the concept. I will demonstrate a twist with the call mapping data set from Section 7.2.2 after the below example is completed.

I have created a very simple data set below called MINSPAN, containing the node and link definitions that include the cost to traverse between each node. The minimum spanning tree algorithm is applied to undirected graphs; so, I have to provide only the From/To for one direction, shorting the definition of my network. If you draw it out on paper, you can see that I have fully defined a 6-node network.

```
data work.minspan;
input From $ To $ Cost;
datalines;
A B 1
A C 3
A E 1
B D 2
B F 4
C D 1
C E 2
D F 1
;
run;
```

Using the above data in PROC OPTNET below, I have defined the necessary FROM, TO, and WEIGHT elements with the DATA_LINKS_VAR statement. And I am invoking the minimum spanning tree algorithm with MINSPANTREE. The resulting data set (the nodes that are part of the MST) is then written to MST_CLASSIC.

```
proc optnet data_links=work.minspan;
    data_links_var from=from to=to weight=cost;
    minspantree out=work.MST_classic;
run;
```

The result of my minimum spanning tree can be viewed using the PROC PRINT statement below, which you can see in Output 7.5.

```
proc print data=work.mst_classic;
run;
```

Output 7.5: Simple Network MST Results

Obs	From	To	Cost
1	D	F	1
2	A	E	1
3	C	D	1
4	A	B	1
5	C	E	2

Now, I want to focus on a different way to use this technique. Sticking with the law enforcement or national security theme established in Section 7.3.1, suppose I want to understand the most influential subnetwork.

Influence is a subjective concept that is usually measured by things like number of connections, emails, calls, etc. It is whatever quantifiable measure for interaction that makes sense in the given context. So, I'm going to apply this idea to the call mapping network generated in Section 7.2.2. I want to find the subnetwork of individuals that are most influential on each other. This is just one way of looking at the notion of influence or information flow in networks, and is based entirely on level of communication between individuals. In this way, I have to assume that if persons A and B call each other daily, and B and C also talk daily, then there is a high likelihood that information from person A will flow to person C (and vice versa).

However, I have to make a change to my call map data set in order to use it in the *minimum* spanning tree. Since I'm looking for those with the highest activity, I have to find the inverse of my call activity counts in order for the minimum search to yield what I need.

```
data work.Inverted;
set work.counts;
Inverse = 1/count;
run;
```

In the DATA step above, I simply create the inverse of my count value from the call mapping result data set COUNTS. That allows me to perform MST on the resulting data set INVERTED to see which individuals created the highest communication channel or influence throughout my network.

```
proc optnet data_links=WORK.Inverted;
    data_links_var from=FromName to=ToName weight=Inverse;
    minspantree out=WORK.MST_links;
run;
proc print data=work.mst_links;
run;
```

The above code runs the MST on the INVERTED data set with the same settings as before, with the output data set created as MST_LINKS. This resulting data set is the printed using PROC PRINT. See the results below.

Output 7.6: MST of Inverse Value Call Map

Obs	FromName	ToName	Inverse
1	LEE	THOMAS	0.16667
2	ROBINSON	WILSON	0.16667
3	RODRIGUEZ	TAYLOR	0.16667
4	CLARK	MOORE	0.16667
5	SMITH	WHITE	0.16667
6	BROWN	GARCIA	0.20000
7	GARCIA	THOMAS	0.20000
8	JONES	THOMPSON	0.20000
9	CLARK	HARRIS	0.20000
10	WILSON	CLARK	0.20000
11	MOORE	RODRIGUEZ	0.20000
12	THOMPSON	ANDERSON	0.20000
13	LEWIS	THOMPSON	0.20000
14	TAYLOR	LEWIS	0.20000
15	ANDERSON	LEE	0.20000
16	DAVIS	MARTINEZ	0.20000
17	MARTINEZ	JACKSON	0.20000
18	THOMAS	WALKER	0.20000
19	WALKER	JACKSON	0.20000
20	MILLER	ROBINSON	0.20000
21	ROBINSON	WHITE	0.20000
22	MARTIN	TAYLOR	0.20000
23	JOHNSON	HARRIS	0.25000
24	WILLIAMS	MARTIN	0.25000

As you can see in Output 7.6 above, the 50 call count pairs that formed my original network of counts has been reduced to only 24 pairs. They are ranked from lowest to highest cost, meaning the top observations actually had the highest call frequencies. So, these are, based on my data, the most influential entities in my network. While the metric validity is possibly debatable, the notion that this is the most communicative subnetwork is not. I say debatable here because such a simple metric may not truly be indicative of influence—additional data may enable you to correlate this metric to the outcome that you want influenced, thus validating or rejecting it in favor of something else. Influence is a tricky thing to measure, but simple metrics about networks can be analyzed within your specific context to assess their utility.

You might be able to imagine at this point that the applications of this approach can quickly become quite complex as I add vectors for connecting these network nodes (i.e., email, texts, etc.); however, I'll let you have fun exploring that on your own.

7.3.3 Clique Detection

Identifying cliques in a network can be another way to identify communities of influence or strong association. For example, marketers like to use clique analysis to perform micro-targeting across social media platforms—if your close friends are doing it, you might do it, too.

Clique: Subnetwork in which every node is connected to every other node in the subgraph.

Going back to the law enforcement domain, you can see that once I've identified a clique in my network, it may either help me see who else is a bad actor, or persons to interview with potential knowledge of a crime.

You can see below that I have pulled my call mapping COUNTS, with the CLIQUE statement used to find cliques within the call data entirely on the undirected connection (i.e., the number of calls doesn't matter here, only that there was a call in one direction between the individuals).

```
proc optnet data_links=WORK.COUNTS loglevel=basic;
    data_links_var from=FromName to=ToName;
    clique out=WORK.MCP_cliques;
run;

proc print data=work.mcp_cliques;
run;
```

After my PROC OPTNET call, you can see the PROC PRINT output for the MCP_CLIQUES data set, generated by OPTNET. Output 7.7 below shows only a sample of the results of PROC PRINT. The number of cliques in my call data set is actually much higher than shown below, but you can see how the approach works. Each person in a clique is shown with their associated clique number (automatically sorted by clique number). So, you can see in Output 7.7 that ANDERSON, LEWIS, and THOMPSON form a clique of size 3, for example.

Output 7.7: Sample of Call Network Cliques

Obs	clique	node
1	1	ANDERSON
2	1	LEWIS
3	1	THOMPSON
4	2	ANDERSON
5	2	LEE
6	3	LEE
7	3	GARCIA
8	3	THOMAS
9	4	LEWIS
10	4	TAYLOR
11	5	BROWN
12	5	GARCIA
13	6	BROWN
14	6	THOMPSON
15	7	GARCIA
16	7	DAVIS

7.3.4 Minimum Cut

Depending on the business context, this can enable you to segment your network into two pieces, comprising the most closely associated individuals based on linkage metrics—but not as close as a clique or biconnected component. Think of these two pieces as communities or subnetworks that you want to identify.

Minimum cut: The minimum cost bisection of the network into two disjoint sets.

Now, like with any optimization routine, you can have multiple solutions to your problem set, and thus ways to generate the segmentation of your network. As a result, you can control the number of solutions PROC OPTNET provides as output. I have chosen to show up to 3 possible solutions by setting the MAXNUMCUTS to 3 in the below code.

```
proc optnet data_links=WORK.COUNTS graph_direction=undirected loglevel=basic
outnodes = MinCuts;
    data_links_var from=FromName to=ToName weight=count;
    mincut
    out=WORK.CallMinCut
    maxnumcuts=3;
run;
```

As with the other analyses I have explored, OPTNET is set to use WORK.COUNTS as an UNDIRECTED data source. And I have invoked the MINCUT statement to tell PROC OPTNET to perform the minimum cut algorithm.

In the below code, I have a PROC PRINT statement for both the MINCUTS and CALLMINCUT data sets, which contain the partition mapping and link sets for the mincut, respectively.

```
proc print data=work.MinCuts;
run;

proc print data=work.callmincut;
run;
```

In Output 7.8 below, you can see there is a binary value associated with each node in my network and every mincut generated. The binary value shows how the network nodes are partitioned (all the 0 nodes are in one partition, while all the 1 nodes are in the other). You can also see that this example doesn't offer a very interesting partition set. This is just an artifact of my phone call data set. As you try different data sources, with different levels of connectedness and weighting, you will see the partitions and mincut sets change dramatically. I would encourage you to play with some sample data sets to get more comfortable with how to use Mincut in the future.

Output 7.8: Node MinCut Matrix

Obs	node	mincut_1	mincut_2	mincut_3
1	ANDERSON	1	1	1
2	LEE	1	1	1
3	LEWIS	1	1	1
4	BROWN	1	1	1
5	GARCIA	1	1	1
6	THOMPSON	1	1	1
7	CLARK	1	1	1
8	HARRIS	1	1	1
9	MOORE	1	1	1
10	DAVIS	1	1	0
11	MARTINEZ	1	1	1
12	THOMAS	1	1	1
13	WILLIAMS	1	0	1
14	JACKSON	1	1	1
15	MILLER	1	1	1
16	SMITH	1	1	1
17	JOHNSON	0	1	1
18	WHITE	1	1	1
19	JONES	1	1	1
20	MARTIN	1	1	1
21	TAYLOR	1	1	1
22	ROBINSON	1	1	1
23	RODRIGUEZ	1	1	1
24	WILSON	1	1	1
25	WALKER	1	1	1

You can see the output for the CALLMINCUT data set below. It contains the link sets broken by the mincut algorithm associated with each mincut generated by PROC OPTNET. Each record includes the weight values for those links as well as the mincut set to which they belong. So, taking the first mincut set as an example, you can see that the JOHNSON node is isolated into its own partition in Output 7.8 above, and the two broken links for JOHNSON are listed in Output 7.9 below.

Output 7.9: Sets of Minimum Cut Pairs

Obs	mincut	FromName	ToName	COUNT
1	1	JOHNSON	HARRIS	4
2	1	JOHNSON	WHITE	2
3	2	HARRIS	WILLIAMS	3
4	2	WILLIAMS	MARTIN	4
5	3	DAVIS	GARCIA	2
6	3	DAVIS	MARTINEZ	5

How you ultimately use this information will depend entirely on your business goals. Suppose you are a marketer that needs to partition the network based on a specific metric to identify communities of interest within your market territory. Alternatively, you may be working to identify ideal targets for breaking up a criminal organization.

7.3.5 Shortest Paths

I will end the overview of analyses with a discussion of shortest paths with PROC OPTNET. There are many applications where you may want to understand the shortest distance (or lowest cost path) between two entities within a network. This determination of distance or cost can be whatever metric makes the most sense for your particular business problem. For example, communications network routing optimization would probably use network latency data as the metric, while a supply chain optimization problem could use financial cost, time, or distance.

Source: The starting node input to a shortest path search algorithm.

Sink: The ending node input to a shortest path algorithm.

For convenience, I'm going to reuse the COUNTS data set I have worked with a bit in this chapter already. Even though the data set is generated to show the counts of calls between individuals, we can imagine it is the cost of traversing that edge instead. I'm again doing this out of convenience as I want to focus on the analysis, and not be so concerned about a new data set.

I'm starting with the code below by showing you how to create all the shortest paths for a network. By specifying only shortest path analysis using the SHORTPATH statement, without identifying source and sink nodes, I will get every permutation of shortest paths for my data set. The resulting paths and path weights are written to SPP_PATHS and SPP_WEIGHTS respectively, which are shown in Output 7.10 and Output 7.11 below.

```
proc optnet data_links=WORK.COUNTS graph_direction=undirected loglevel=basic;
    data_links_var from=FromName to=ToName weight=COUNT;
    shortpath
    out_paths=WORK.SPP_Paths
    out_weights=WORK.SPP_Weights;
run;

proc print data=work.spp_paths;
run;

proc print data=work.spp_weights;
run;
```

The shortest path permutations for the COUNTS data set results in 1,962 records, of which I show only a sample in Output 7.10 below (WORK.SPP_PATHS). This data set contains a record for every segment of the source, sink path.

Output 7.10: All Shortest Paths

Obs	source	sink	order	FromName	ToName	COUNT
1	ANDERSON	LEE	1	ANDERSON	LEE	5
2	ANDERSON	LEWIS	1	ANDERSON	LEWIS	2
3	ANDERSON	BROWN	1	ANDERSON	THOMPSON	5
4	ANDERSON	BROWN	2	THOMPSON	BROWN	2
5	ANDERSON	GARCIA	1	ANDERSON	LEE	5
6	ANDERSON	GARCIA	2	LEE	GARCIA	5
7	ANDERSON	THOMPSON	1	ANDERSON	THOMPSON	5
8	ANDERSON	CLARK	1	ANDERSON	LEWIS	2
9	ANDERSON	CLARK	2	LEWIS	TAYLOR	5
10	ANDERSON	CLARK	3	TAYLOR	RODRIGUEZ	6
11	ANDERSON	CLARK	4	RODRIGUEZ	MOORE	5
12	ANDERSON	CLARK	5	MOORE	CLARK	6
13	ANDERSON	HARRIS	1	ANDERSON	LEWIS	2
14	ANDERSON	HARRIS	2	LEWIS	TAYLOR	5
15	ANDERSON	HARRIS	3	TAYLOR	MARTIN	5
16	ANDERSON	HARRIS	4	MARTIN	WILLIAMS	4
17	ANDERSON	HARRIS	5	WILLIAMS	HARRIS	3
18	ANDERSON	MOORE	1	ANDERSON	LEWIS	2
19	ANDERSON	MOORE	2	LEWIS	TAYLOR	5
20	ANDERSON	MOORE	3	TAYLOR	RODRIGUEZ	6
21	ANDERSON	MOORE	4	RODRIGUEZ	MOORE	5
22	ANDERSON	DAVIS	1	ANDERSON	LEE	5
23	ANDERSON	DAVIS	2	LEE	GARCIA	5
24	ANDERSON	DAVIS	3	GARCIA	DAVIS	2
25	ANDERSON	MARTINEZ	1	ANDERSON	LEE	5

Below, Output 7.11 is a sample of the shortest path weight data set (WORK.SPP_WEIGHTS), containing 600 records. Notice that this output shows just one record for every source, sink pair as it is the total weight for that entire path.

Output 7.11: Path Weight Totals

Obs	source	sink	path_weight
1	ANDERSON	LEE	5
2	ANDERSON	LEWIS	2
3	ANDERSON	BROWN	7
4	ANDERSON	GARCIA	10
5	ANDERSON	THOMPSON	5
6	ANDERSON	CLARK	24
7	ANDERSON	HARRIS	19
8	ANDERSON	MOORE	18
9	ANDERSON	DAVIS	12
10	ANDERSON	MARTINEZ	17
11	ANDERSON	THOMAS	11
12	ANDERSON	WILLIAMS	16
13	ANDERSON	JACKSON	21
14	ANDERSON	MILLER	20
15	ANDERSON	SMITH	24
16	ANDERSON	JOHNSON	23
17	ANDERSON	WHITE	25
18	ANDERSON	JONES	10
19	ANDERSON	MARTIN	12
20	ANDERSON	TAYLOR	7
21	ANDERSON	ROBINSON	25
22	ANDERSON	RODRIGUEZ	13
23	ANDERSON	WILSON	29
24	ANDERSON	WALKER	16
25	LEE	ANDERSON	5
26	LEE	LEWIS	7

Now that I have reviewed how to find all the shortest paths within the source data set, COUNTS, I want to show you how to make a couple of minor tweaks to the code to specify a single source and sink.

```
proc optnet data_links=WORK.COUNTS graph_direction=undirected loglevel=basic;
    data_links_var from=FromName to=ToName weight=COUNT;
    shortpath
            source=ANDERSON
            sink=CLARK
    out_paths=WORK.SourceSink_Paths
    out_weights=WORK.SourceSink_Weights;
run;
```

Notice in the above code that all I have to do is add the options, SOURCE= and SINK= after the SHORTPATH statement. Further notice that the text information that I put into the options is not surrounded by quotation marks, but simply typed in. I have identified my source as ANDERSON and my sink value as CLARK. You can see the difference in Output 7.12 below.

```
proc print data=work.sourcesink_paths;
run;

proc print data=work.sourcesink_weights;
run;
```

Output 7.12: Path and Weight Results

Obs	source	sink	order	FromName	ToName	COUNT
1	ANDERSON	CLARK	1	ANDERSON	LEWIS	2
2	ANDERSON	CLARK	2	LEWIS	TAYLOR	5
3	ANDERSON	CLARK	3	TAYLOR	RODRIGUEZ	6
4	ANDERSON	CLARK	4	RODRIGUEZ	MOORE	5
5	ANDERSON	CLARK	5	MOORE	CLARK	6

Obs	source	sink	path_weight
1	ANDERSON	CLARK	24

Output 7.12 above shows the shortest path between Anderson and Clark, with all the segments in the first PROC PRINT result and the total path weight in the second PROC PRINT result.

7.4 Summary

In this chapter, I walked through methods for mapping networks of entities based on shared characteristics or interactions, and demonstrated a variety of analyses that you can perform on the resulting maps. I hope you found this enlightening and motivating, and a practical assistance to your work.

Keep in mind that there are many different approaches to the same problem in SAS. And I hope you experiment with different variations on the methods and approaches I have shown thus far.

I am going to wrap up in the next chapter by focusing on entity management, the final phase of the ERA process. The focus on entity management will be less technical than the information covered thus far. Instead, I will concentrate on the policy and business aspects to effectively managing your entity reference data.

---

[1] SAS Institute Inc., *SAS/OR® 14.3 User's Guide: Network Optimization Algorithms* (Cary, NC: SAS Institute Inc.: 2017), 106–107, http://support.sas.com/documentation/onlinedoc/or/143/ornoaug.pdf.

Chapter 8: Entity Management

8.1 Introduction

It is important for every organization to maintain an authoritative repository of entity references, regardless of the end use. Such a repository allows organizations to compare incoming entity references against an existing, authoritative source. So, in this chapter, I'm going to discuss some best practices to ensure that you are thinking about how to build and maintain a clean entity repository.

Figure 8.1: ERA Flow with Entity Management Focus

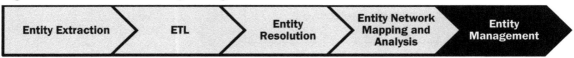

A repository facilitates the ability of an organization to avoid a host of problems such as these:

- duplicate customer accounts
- incorrect invoicing
- improperly sharing personally identifiable information (PII)
- identity theft
- inability to perform entity-level analytics

When you "see" an entity reference for the first time, you know very little about it compared to an entity that is known to your organization. So, when thinking about reference management, you have to think about the primary aspects of the vetting process with the business context in mind.

Specifically, what is your business process trying to achieve? What do you need to *do* with this entire workflow? For instance, when applying the end-to-end ERA process in a retail customer relationship management setting as compared to a law enforcement intelligence gathering setting, the decision criteria for inclusion or exclusion of entity references will be very different. In addition, the criteria for making updates to the entity database will differ by specific business needs.

I have outlined a generic workflow for managing entity references, regardless of business context, in Figure 8.2. The goal here is really to provide guidelines for how to approach entity management rather than a prescriptive process, as the business context makes the specific needs vary.

However, I will walk through these elements as a means of helping you think through the details of this process.

Figure 8.2: Entity Management Workflow

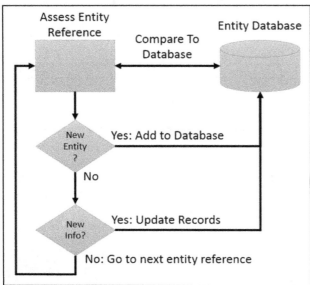

The first step in the process is to assess the entity reference to determine whether it even has enough information to be reviewed for incorporation in the entity database. I anticipate that your work in this vein is accompanied by the planning documentation discussed in Chapter 3, because that will contain most of the information that you need at this stage to assess entity references—*filtering out the ones that don't even contain enough of your key criteria for consideration.*

Next, use the entity database shown in Figure 8.2 as the gold standard for determining the surviving reference (the concept of surviving references was discussed in Section 6.3). If the precise entity reference being assessed already exists in your database, the database record will remain unchanged. Keep in mind that "precise" is based on your predefined match criteria. However, if the entity reference is substantively different from what already exists in the database, a new record is created using the newly identified reference.

If the newly discovered reference is for a currently tracked entity, but contains new information, you may need to update the existing database records. This will depend on the documented decisions regarding matching thresholds in your database. And the thresholds will be driven by the risk preferences of your specific context.

As always, the criteria for each decision go back to the context in which you are working. The context of your application will determine the entity matching thresholds, error tolerances, and criteria for adding or updating the entity references.

Example: A customer relationship management system within a large company determines whether an incoming order is from either a new or existing customer. If it is a new customer, then a new identity record is created in the system. Otherwise, the order information is associated with an existing identity.

In the CRM system example above, the company must protect personally identifiable information (PII) at the possible risk of creating duplicate entity records. The potential risk to both the customer and the company of *duplicate* records is much lower than that of accidental sharing of PII. I will discuss more about the criteria for creating new records in the next section.

The same is true for possibly editing the current customer record, as you wouldn't want to accidentally update the records for entities that are similar, but actually different in reality. I will discuss that further in Section 8.3.

8.2 Creating New Records

Establishing a new entity reference in your system of record requires a high level of certainty in order to avoid duplicate records. So, you need to create a records management plan—if you don't already have one—that identifies the specific elements of each entity reference that must be "new" in order to create a new entry in your entity database. And you must ensure through rigorous testing of your source data that new entries are not generated erroneously.

So, as part of the records management plan, you should have lists of data elements by the sensitivity of information that they contain, and by how frequently they change. The sensitivity aspect will inform your data handling policies (outside the scope of this text), while the frequency of information changes will drive which elements you can depend on for your gold standard entity reference data (i.e., what you will match on).

Depending on your context, you should have several elements of an entity that never change:

> Social Security number (SSN)
> date of birth (DOB)
> place of birth (POB)
> gender at birth

Simultaneously, there are elements of an entity that may change, but relatively infrequently or rarely:

> full name
> marital status

And finally, there are other elements that are likely to change much more frequently:

> address
> phone number
> credit card number

By establishing a plan for which elements never change in your context, you can ensure that the entity reference being evaluated will be accurate with a high level of certainty. If it is not possible to depend entirely on elements that never change, then some combination of elements that never change with elements that change very rarely should provide a robust set of thresholds. The final determination for the best approach should be based on testing your plan on the available database. Regardless, you should avoid using the elements that change frequently, as it is too likely to generate new entities erroneously.

8.3 Editing Existing Records

Depending on your business context, making updates to existing entities requires an even higher level of confidence to certify the entity reference. Why? Well, under many circumstances, it would be worse to edit an existing entity reference entry in the database than to accidentally add a new entry.

Example: Similar, but different, customer records get mixed up. The potential impacts include sharing of PII, or incorrectly addressed items. Depending upon the scale of the mistake, this could lead to catastrophic impact to your business.

So, again, you have to go back to your risk tolerance for the business context at hand, and establish rules and processes to protect existing records from being updated based on your risk attitude.

Accordingly, if the resulting threshold for existing record matches is not achieved, absolutely no automated updates should occur. At that point, a human would need to intervene in order to force the system updates to occur. Appropriate logging of their activity should also be enforced under such circumstances. An emphasis on protecting entity reference data cannot be overstated here, which is why you should always take redundant measures to protect user privacy and avoid corrupting your data stores.

Always refer back to a records management plan, or system of record documentation, to ensure you are appropriately following your institution's policies. These policies have both practical and legal implications for your organization, and should be adhered to as closely as possible. When they are unclear, work with management to get clarification, and document that clarification for future reference.

If such documents do not exist, work with your management to create them. And realize these documents do not necessarily need to conform to a specific format. It is more important that they clearly communicate to the recipient the information needed to establish clear policies and processes, and how said reader can follow them. Due to the practical and legal implications, you will generally have some foundational framework within which you have to work for your industry. So, you should have a starting point, and a control board to review your documentation. However, whatever you put together must be clear for the eventual reader. If a policy or process for managing your entity reference database is unclear for readers, they will possibly make the wrong decisions with a document that has been thoroughly vetted. Effective documentation is worth the effort.

8.4 Summary

We have finally arrived at the end of the book. This chapter was a brief dissection of the key elements that you need to be thinking about for managing entities in your business context. I hope this has proven a useful introduction to the concepts for entity resolution analytics, as well as provided you some new tools for tackling a diverse set of problems in your work.

We have come a long way from just getting started with regular expressions, to now applying some relatively powerful analytics with concepts built on that foundation. Each topic area that we have explored in this book really could have been its own book. I hope you walk away equipped to handle these different topics, while appreciating that there is so much more to learn in each. And I hope you use this introduction as a jumping off point to dig deeper into whatever areas you found most intriguing.

Thank you for reading, and happy coding!

Appendix A: Additional Resources

A.1 Perl Version Notes

This section contains notes about the limitations of the Perl version that is used by SAS 9.4.[1] It is intended for the Perl experts out there who are wondering about some of the missing pieces within SAS. The information below was taken directly from the SAS website and is provided here with some slight modifications because it is somewhat difficult information to find but is potentially very useful to the advanced reader of this book.

The PRX functions use a modified version of Perl 5.6.1 to perform regular expression compilation and matching. Perl is compiled into a library for use with SAS. This library is shipped with SAS® 9. The modified and original Perl 5.6.1 files are freely available in a ZIP file from the Technical Support Web site. The ZIP file is provided to comply with the Perl Artistic License and is not required in order to use the PRX functions. Each of the modified files has a comment block at the top of the file describing how and when the file was changed. The executables were given nonstandard Perl names. The standard version of Perl can be obtained from the Perl Web site.

Only Perl regular expressions are accessible from the PRX functions. Other parts of the Perl language are not accessible. The modified version of Perl [RegEx] does not support the following items:

- All Perl variables, except for the capture buffer variables $1 - $n
- Metacharacters \G, \pP, \PP, and \X
- RegEx options /c and /g, and /e with substitutions
- Named characters (i.e., \N{name})
- Executing Perl code within a regular expression, which includes the syntax (?{code}), (??{code}), and (?p{code})
- Unicode pattern matching
- Pattern delimiters other than the backslash. For example: ?PATTERN?, !PATTERN!, etc.
- Perl code comments between a pattern and replacement text. For example: s{regexp} #perl comment {replacement text}
- Using matching backslashes with m/\\\\/ instead of m/\\/ to match a backslash[2]

A.2 ASCII Code Lookup Tables

A.2.1 Non-Printing Characters

Binary	Hex	Dec	Oct	ASCII Abbr.	Crtl Character (Command Prompt Display)	Description
0000 0000	00	0	000	NUL	^@	Null Character
0000 0001	01	1	001	SOH	^A	Start of Header
0000 0010	02	2	002	STX	^B	Start of Text
0000 0011	03	3	003	ETX	^C	End of Text
0000 0100	04	4	004	EOT	^D	End of Transmission
0000 0101	05	5	005	ENQ	^E	Enquiry
0000 0110	06	6	006	ACK	^F	Acknowledgment
0000 0111	07	7	007	BEL	^G	Bell
0000 1000	08	8	010	BS	^H	Backspace
0000 1001	09	9	011	HT	^I	Horizontal Tab
0000 1010	0A	10	012	LF	^J	Line Feed
0000 1011	0B	11	013	VT	^K	Vertical Tab
0000 1100	0C	12	014	FF	^L	Form Feed
0000 1101	0D	13	015	CR	^M	Carriage Return
0000 1110	0E	14	016	SO	^N	Shift Out
0000 1111	0F	15	017	SI	^O	Shift In
0001 0000	10	16	020	DLE	^P	Data Link Escape
0001 0001	11	17	021	DC1	^Q	Device Control 1 (oft. XON)
0001 0010	12	18	022	DC2	^R	Device Control 2
0001 0011	13	19	023	DC3	^S	Device Control 3 (oft. XOFF)
0001 0100	14	20	024	DC4	^T	Device Control 4
0001 0101	15	21	025	NAK	^U	Negative Acknowledgment
0001 0110	16	22	026	SYN	^V	Synchronous Idle

Binary	Hex	Dec	Oct	ASCII Abbr.	Crtl Character (Command Prompt Display)	Description
0001 0111	17	23	027	ETB	^W	End of Transmission Block
0001 1000	18	24	030	CAN	^X	Cancel
0001 1001	19	25	031	EM	^Y	End of Medium
0001 1010	1A	26	032	SUB	^Z	Substitute
0001 1011	1B	27	033	ESC	^[	Escape
0001 1100	1C	28	034	FS	^\	File Separator
0001 1101	1D	29	035	GS	^]	Group Separator
0001 1110	1E	30	036	RS	^^[j]	Record Separator
0001 1111	1F	31	037	US	^_	Unit Separator
0111 1111	7F	127	177	DEL	^?	Delete

A.2.2 Printing Characters

Binary	Hex	Dec	Oct	Display Character
0010 0000	20	32	040	
0010 0001	21	33	041	!
0010 0010	22	34	042	"
0010 0011	23	35	043	#
0010 0100	24	36	044	$
0010 0101	25	37	045	%
0010 0110	26	38	046	&
0010 0111	27	39	047	'
0010 1000	28	40	050	(
0010 1001	29	41	051	)

Binary	Hex	Dec	Oct	Display Character
0010 1010	2A	42	052	*
0010 1011	2B	43	053	+
0010 1100	2C	44	054	,
0010 1101	2D	45	055	-
0010 1110	2E	46	056	.
0010 1111	2F	47	057	/
0011 0000	30	48	060	0
0011 0001	31	49	061	1
0011 0010	32	50	062	2
0011 0011	33	51	063	3
0011 0100	34	52	064	4
0011 0101	35	53	065	5
0011 0110	36	54	066	6
0011 0111	37	55	067	7
0011 1000	38	56	070	8
0011 1001	39	57	071	9
0011 1010	3A	58	072	:
0011 1011	3B	59	073	;
0011 1100	3C	60	074	<
0011 1101	3D	61	075	=
0011 1110	3E	62	076	>
0011 1111	3F	63	077	?
0100 0000	40	64	100	@
0100 0001	41	65	101	A
0100 0010	42	66	102	B

Binary	Hex	Dec	Oct	Display Character
0100 0011	43	67	103	C
0100 0100	44	68	104	D
0100 0101	45	69	105	E
0100 0110	46	70	106	F
0100 0111	47	71	107	G
0100 1000	48	72	110	H
0100 1001	49	73	111	I
0100 1010	4A	74	112	J
0100 1011	4B	75	113	K
0100 1100	4C	76	114	L
0100 1101	4D	77	115	M
0100 1110	4E	78	116	N
0100 1111	4F	79	117	O
0101 0000	50	80	120	P
0101 0001	51	81	121	Q
0101 0010	52	82	122	R
0101 0011	53	83	123	S
0101 0100	54	84	124	T
0101 0101	55	85	125	U
0101 0110	56	86	126	V
0101 0111	57	87	127	W
0101 1000	58	88	130	X
0101 1001	59	89	131	Y

Binary	Hex	Dec	Oct	Display Character
0101 1010	5A	90	132	Z
0101 1011	5B	91	133	[
0101 1100	5C	92	134	\
0101 1101	5D	93	135	]
0101 1110	5E	94	136	^
0101 1111	5F	95	137	_
0110 0000	60	96	140	`
0110 0001	61	97	141	a
0110 0010	62	98	142	b
0110 0011	63	99	143	c
0110 0100	64	100	144	d
0110 0101	65	101	145	e
0110 0110	66	102	146	f
0110 0111	67	103	147	g
0110 1000	68	104	150	h
0110 1001	69	105	151	i
0110 1010	6A	106	152	j
0110 1011	6B	107	153	k
0110 1100	6C	108	154	l
0110 1101	6D	109	155	m
0110 1110	6E	110	156	n
0110 1111	6F	111	157	o
0111 0000	70	112	160	p

Binary	Hex	Dec	Oct	Display Character
0111 0001	71	113	161	q
0111 0010	72	114	162	r
0111 0011	73	115	163	s
0111 0100	74	116	164	t
0111 0101	75	117	165	u
0111 0110	76	118	166	v
0111 0111	77	119	167	w
0111 1000	78	120	170	x
0111 1001	79	121	171	y
0111 1010	7A	122	172	z
0111 1011	7B	123	173	{
0111 1100	7C	124	174	|
0111 1101	7D	125	175	}
0111 1110	7E	126	176	~

A.3 POSIX Metacharacters

Throughout the book, we discussed metacharacters of all types that adhere to Perl standards (de facto standard across the industry) for implementation since they are what SAS uses. And they are all that you need when you're running within the SAS environment. However, if you ever need to push the RegEx processing to a system outside of SAS, there is no guarantee that they will always work because not all systems use Perl syntax (mostly older systems don't).

Note: When you are attempting this more advanced application, know the parameters of the system you are using. You might not need to change the RegEx coding.

The exact applications of the metacharacters described in this section are outside the scope of this text but are provided here for the advanced reader who is interested in them. For example, although we have not covered it, POSIX metacharacters might be needed when you are performing in-database *fuzzy matching* with PROC SQL.

[[:alpha:]]
> This metacharacter matches any alphabetic character and is equivalent to [a-zA-Z].

[[:^alpha:]]
> This metacharacter matches any non-alphabetic character and is equivalent to [^a-zA-Z].

[[:alnum:]]
> This metacharacter matches any alphanumeric character and is equivalent to [a-zA-Z0-9].

[[:^alnum:]]
> This metacharacter matches any non-alphanumeric character and is equivalent to [^a-zA-Z0-9].

[[:ascii:]]
> This metacharacter matches any ASCII character and is equivalent to [\0-\177] (i.e., it does not match UNICODE).

[[:^ascii:]]
> This metacharacter matches any non-ASCII character and is equivalent to [^\0-\177] (i.e., it matches UNICODE).

[[:blank:]]
> This metacharacter matches any blank character.

[[:^blank:]]
> This metacharacter matches any non-blank character.

[[:cntrl:]]
> This metacharacter matches any control character.

[[:^cntrl:]]
> This metacharacter matches any non-control character.

[[:digit:]]
> This metacharacter matches any digit character and is equivalent to \d or [0-9].

[[:^digit:]]
> This metacharacter matches any non-digit character and is equivalent to \D and [^0-9].

[[:graph:]]
> This metacharacter matches any visible character and is equivalent to [[:alnum:][:punct:]]. In other words, if you can see it when printed on a piece of paper, then it is matched by this metacharacter.

[[:^graph:]]
> This metacharacter matches any non-printing character and is equivalent to [^[:alnum:][:punct:]]. If you can't see it printed on a piece of paper, then it is matched by this metacharacter.

[[:lower:]]
> This metacharacter matches any lowercase alphabetic character and is equivalent to [a-z].

[[:^lower:]]
> This metacharacter matches anything except a lowercase alphabetic character and is equivalent to [^a-z].

[[:print:]]

    This metacharacter prints a string of characters—any characters encountered.

[[:^print:]]

    This metacharacter does not print any characters.

[[:punct:]]

    This metacharacter matches any visible punctuation or symbol character.

[[:^punct:]]

    This metacharacter matches anything except visible punctuation or symbol characters.

[[:space:]]

    This metacharacter matches any space character and is equivalent to \s.

[[:^space:]]

    This metacharacter matches anything except a space character and is equivalent to \S.

[[:upper:]]

    This metacharacter matches any uppercase alphabetic characters and is equivalent to [A-Z].

[[:^upper:]]

    This metacharacter matches all non-uppercase alphabetic characters and is equivalent to [^A-Z].

[[:word:]]

    This metacharacter matches any word character encountered and is equivalent to \w.

[[:^word:]]

    This metacharacter matches any non-word characters and is equivalent to \W.

[[:xdigit:]]

    This metacharacter matches any hexadecimal character.

[[:^xdigit:]]

    This metacharacter does not match a hexadecimal character.

A.4 Random PII Generation

The output from the code discussed in this section is used throughout the book. To avoid the distraction of your encountering it embedded in the chapters, I have moved it to the appendix for you to study outside the flow of the book. I hope you find this a sensible approach, and easy to follow. It is best to review this section, and run the included code, starting in Chapter 4, Entity Extraction, as that is where I introduce the output for those examples.

A.4.1 Random PII Generator

The following code was developed to provide a set number of randomly generated elements for the purposes of the following examples. This process is an effort to replicate the kind of data we all see on a regular basis, PII, without encountering the usual privacy issues associated with it. As we will see in the code, every effort was made to make these elements feel real. However, it is worth noting that more advanced techniques (and more efficient techniques) were not employed because of the introductory nature of this text. If you're interested, use this code as a baseline to develop a more sophisticated and efficient random PII generator. Doing so is a great way to support both learning and real-world development work.

Note: All occurrences of PII shown in the coming pages were generated in a random fashion. Any resemblance to actual PII is completely coincidental.

A static snapshot of randomly generated data is used below, and it is not guaranteed to be replicated. But the parameters for any data set that is created by this code will be the same for any data set. Also, the code uses a few different methods for creating the various data elements. I want to demonstrate the variety of methods available to us for doing any task in SAS.

Much of this code is unavoidably long due to the steps taken to create names, addresses, and other information. However, I hope it is commented well enough for you to follow. It should be easy to understand, and update to your needs, after you have run through it a few times.

```
/*First, we create data sets for First and Last names*/
data FirstNames;  ❶
input Firstname $20.;
/*Common First Names (male and female) in the United States*/
datalines;
JAMES
JOHN
ROBERT
MICHAEL
WILLIAM
DAVID
RICHARD
CHARLES
JOSEPH
THOMAS
CHRISTOPHER
DANIEL
PAUL
MARK
DONALD
GEORGE
KENNETH
STEVEN
EDWARD
BRIAN
RONALD
ANTHONY
KEVIN
JASON
MATTHEW
MARY
PATRICIA
LINDA
BARBARA
ELIZABETH
JENNIFER
MARIA
SUSAN
MARGARET
DOROTHY
LISA
```

```
NANCY
KAREN
BETTY
HELEN
SANDRA
DONNA
CAROL
RUTH
SHARON
MICHELLE
LAURA
SARAH
KIMBERLY
DEBORAH
;
run;

data surnames;
input Surname $20.;
/*Common Last Names in the United States*/
datalines;
SMITH
JOHNSON
WILLIAMS
JONES
BROWN
DAVIS
MILLER
WILSON
MOORE
TAYLOR
ANDERSON
THOMAS
JACKSON
WHITE
HARRIS
MARTIN
THOMPSON
GARCIA
MARTINEZ
ROBINSON
CLARK
RODRIGUEZ
LEWIS
LEE
WALKER
;
run;

/*Next, we take a simple random sample of a fixed number of names*/
proc surveyselect data=firstnames method=srs n=25      ❷
                  out=firstnamesSRS;
run;
proc surveyselect data=surnames method=srs n=25
                  out=surnamesSRS;
run;
```

```
/*We must create an index value to perform match-merge on later*/
data firstnamesSRS;
set firstnamesSRS;
    Num=_N_;
run;
data surnamesSRS;
set surnamesSRS;
    Num=_N_;
run;

data PII_Numbers;
n = 25; /*Determines the number of records we will create.*/

/*Arrays used for random day creation below*/❸
array x x1-x12 (1:12);
array d d1-d28 (1:28);
array a a1-a30 (1:30);
array y y1-y31 (1:31);
array z z1-z20 (1974:1994);
seed=1234567890123; /*Random Number Seed Value*/

do i= 1 to n; /*Master Loop for PhoneNumber, Date of Birth, and SSN*/
    Num=i; /*Num is used as a unique index value for data set merging later*/
    /*First, we randomly create the number segments*/❹
    CountryCode = Strip(INT(10*rand('UNIFORM')));
    AreaCode =
Compress(INT(10*rand('UNIFORM'))||INT(10*rand('UNIFORM'))||INT(10*rand('UNIFORM')
));
    NextThree =
Compress(INT(10*rand('UNIFORM'))||INT(10*rand('UNIFORM'))||INT(10*rand('UNIFORM')
));
    LastFour =
Compress(INT(10*rand('UNIFORM'))||INT(10*rand('UNIFORM'))||INT(10*rand('UNIFORM')
)||INT(10*rand('UNIFORM')));

    /*Next, we randomly create common separator types*/
    separator = rand('UNIFORM');❺
    if separator >= .66 then do;
            PhoneNumber = Compress(CountryCode||'-'||AreaCode||'-'||NextThree||'-
'||LastFour);
    end;
    else if separator >=.33 AND separator <.66 then do;
            PhoneNumber = Compress(CountryCode||'
'||'('||AreaCode||')'||NextThree||'-'||LastFour);
    end;
    else if separator <.33 then do;
            PhoneNumber =
Compress('+'||CountryCode||'.'||AreaCode||'.'||NextThree||'.'||LastFour);
    end;

    /*Social Security Number*/  ❻
    SSN=Compress(INT(10*rand('UNIFORM'))||INT(10*rand('UNIFORM'))||INT(10*rand('UN
IFORM'))||'-'||
            INT(10*rand('UNIFORM'))||INT(10*rand('UNIFORM'))||'-'||

    INT(10*rand('UNIFORM'))||INT(10*rand('UNIFORM'))||INT(10*rand('UNIFORM'))||INT
(10*rand('UNIFORM')));
```

```
    /*Date Of Birth*/
      call ranperk(seed, 1, of x1-x12);        ❼
     month=x1;
     if x1=2 then do;
          call ranperk(seed, 1, of d1-d28);
          day=d1;
          end;
     else if x1 in (4,6,9,11) then do;
          call ranperk(seed, 1, of a1-a30);
          day=a1;
          end;
     else if (x1=1|x1=3|x1=5|x1=7|x1=8|x1=10|x1=12) then do;
          call ranperk(seed, 1, of y1-y31);
          day=y1;
          end;
     call ranperk(seed, 1, of z1-z20);
     year=z1;
    DOB=compress(month||'/'||day||'/'||year);

    output; /*OUTPUT must be made explicit within a DO LOOP*/
end;                /*The DATA step only runs once because there is no data.*/

keep Num SSN DOB PhoneNumber; /*The only elements we need for the next step*/
run;

/*Now we extract the addresses from a file using RegEx*/
data Addresses;
infile 'F:\Unstructured Data Analysis\Appendix_A_Example_Source\addresses.txt'
length=linelen lrecl=500 pad;
varlen=linelen-0;

input source_text $varying500. varlen; ❽
pattern = "/^(\d+?)\t(.+)/o";
pattern_ID = prxparse(pattern);
position = PRXMATCH(pattern_ID, source_text);

if PRXMATCH(pattern_ID, source_text) then do;
   Num = PRXPOSN(pattern_ID, 1, source_text) * 1;
   Address = PRXPOSN(pattern_ID, 2, source_text);
end;

keep Num Address;
run;

proc print data=addresses;
run;

/*Now, we create the PII data set with match-merge*/   ❾
data PII;
merge firstnamesSRS surnamesSRS PII_Numbers addresses;
by num;
drop num;
run;

proc print data=PII;
run;
```

❶ We start with an easy way to create pseudo-random names, by just creating name data sets using data lines. It is not elegant or short, but it gets the job done for our purposes.

❷ Here we are sampling the name data sets, using simple random sampling and a sample size of n=25. The sample size is completely arbitrary and chosen to match the number of other random values created later in the code.

❸ The arrays are created to ensure that legitimate date values can be created for our arbitrary range of years. To avoid any complications, we are ignoring leap years (no Feb 29th in the set of possibilities) and are using an arbitrary set of 4-digit years. The seed value is an arbitrary number.

❹ Now, we construct the phone number by using the RAND function (UNIFORM option) to generate the individual digits. The INT function takes the integer portion of a value, so multiplying the random value between 0 and 1 by 10 and applying the INT function yields a single digit between 0 and 9. This method ensures that zero values are not dropped (a leading zero would otherwise not be held). Other methods can achieve the same outcome, but this is a straightforward implementation without the need for arrays. The COMPRESS function is used to remove all spaces between the connected values. However, removing this function is an easy way to make the data messier.

❺ After creating the individual chunks of a phone number, we randomly assign different separator types in an effort to demonstrate the various representations that might be expected in practice.

❻ Next, we create Social Security numbers (SSNs) by applying the same techniques as with the phone numbers immediately above. However, we are not randomizing the separator. It is less often an issue, but you could do it as an extracurricular exercise.

❼ We now build the date of birth (DOB) using the arrays discussed in ❸ and the RANPERK function. This function creates random permutations of the provided arrays and provides k values from the results. Other methods could have been employed, but this is a simple approach to create random date elements within a specific range (i.e., valid dates).

❽ The DATA step for addresses uses some familiar RegEx functionality to extract addresses from a text file, along with the Num value that allows us to perform a match-merge in the next step.

❾ This final DATA step creates a single data set, PII, from the above elements.

A.4.2 Output

Output A.1 displays the final data set created by our code, Rand_PII_Generator.sas. As expected, it contains 25 pseudo-random PII elements to support some of our upcoming examples.

Output A.1: Rand_PII_Generator.sas Sample Output

Obs	Firstname	Surname	PhoneNumber	SSN	DOB	Address
1	JAMES	SMITH	2-746-475-4589	539-71-9216	12/17/1986	1776 D St NW, Washington, DC 20006
2	JOHN	JOHNSON	0(439)270-9250	189-03-1020	3/25/1981	1600 Pennsylvania Ave NW, Washington, DC 20500
3	WILLIAM	WILLIAMS	+6.281.794.3626	971-45-0631	10/2/1986	600 14th St NW, Washington, DC 20005
4	DAVID	JONES	9(349)208-5935	277-05-1098	8/9/1985	1321 Pennsylvania Ave NW, Washington, DC 20004
5	THOMAS	BROWN	3(287)870-2874	123-22-9494	1/26/1980	2470 Rayburn Hob, Washington, DC 20515
6	CHRISTOPHER	DAVIS	6(639)100-7721	688-49-1392	5/20/1992	101 Independence Ave SE, Washington, DC 20540
7	DANIEL	MILLER	8(277)323-9564	675-36-2461	9/9/1977	511 10th St NW, Washington, DC 20004
8	PAUL	WILSON	6-863-034-6857	304-59-8869	5/18/1980	450 7th St NW, Washington, DC 20004
9	MARK	MOORE	5(697)801-1886	102-01-9574	7/29/1991	2 15th St NW, Washington, DC 20007
10	DONALD	TAYLOR	3(019)416-1550	439-72-6850	2/1/1982	3700 O St NW, Washington, DC 20057
11	GEORGE	ANDERSON	0(036)200-7891	116-12-3837	8/10/1983	3001 Connecticut Ave NW, Washington, DC 20008
12	KENNETH	THOMAS	1(705)466-8443	549-54-0433	5/22/1976	3101 Wisconsin Ave NW, Washington, DC 20016
13	EDWARD	JACKSON	+7.841.664.8908	461-17-1160	9/6/1977	800 Florida Ave NE, Washington, DC 20002
14	ANTHONY	WHITE	8-451-939-9401	374-43-5208	5/7/1979	1 First St NE, Washington, DC 20543
15	KEVIN	HARRIS	7-690-620-4418	877-45-2254	6/10/1984	600 Independence Ave SW, Washington, DC 20560
16	MATTHEW	MARTIN	+2.623.941.6074	436-07-9380	10/29/1975	10th St. & Constitution Ave. NW, Washington, DC 20560
17	LINDA	THOMPSON	6-564-897-1662	500-98-4809	3/10/1983	555 Pennsylvania Ave NW, Washington, DC 20001
18	MARGARET	GARCIA	5(894)411-7166	772-03-0744	10/16/1992	1000 5th Ave, New York, NY 10028
19	LISA	MARTINEZ	+1.167.511.7529	426-25-2712	8/18/1978	64th St and 5th Ave, New York, NY 10021
20	KAREN	ROBINSON	5-395-540-6931	761-09-3862	10/18/1974	10 Lincoln Center Plaza, New York, NY 10023
21	DONNA	CLARK	7-781-276-1544	131-47-7120	11/3/1991	Pier 86 W 46th St and 12th Ave, New York, NY 10036
22	CAROL	RODRIGUEZ	9(737)785-2677	280-95-9464	12/28/1985	350 5th Ave, New York, NY 10118
23	SHARON	LEWIS	0(918)040-2361	896-33-5968	1/31/1975	405 Lexington Ave, New York, NY 10174
24	MICHELLE	LEE	9(488)783-8608	735-62-9285	10/3/1984	1 Albany St, New York, NY 10006
25	KIMBERLY	WALKER	7-823-096-2389	109-02-1649	9/17/1990	1585 Broadway, New York, NY 10036

---

[1] The available Perl RegEx functionality has not changed since SAS 9.1. These notes are current as of the writing of this book. For the most up-to-date information regarding versioning, please visit the SAS documentation website at: http://support.sas.com/documentation/.

[2] SAS Institute Inc. "Perl Artistic License Compliance," *SAS® 9.4 Functions and CALL Routines: Reference, Fifth Edition,* http://support.sas.com/documentation/cdl/en/lefunctionsref/67239/HTML/default/viewer.htm#p0tw80fkq qpow5n1f7xwvd6bsonq.htm (accessed August 29, 2018).

Ready to take your SAS® and JMP® skills up a notch?

Be among the first to know about new books, special events, and exclusive discounts.
support.sas.com/newbooks

Share your expertise. Write a book with SAS.
support.sas.com/publish

sas.com/books
for additional books and resources.

§.sas.
THE POWER TO KNOW.

www.ingramcontent.com/pod-product-compliance
Lightning Source LLC
Chambersburg PA
CBHW060111090326
40690CB00064B/5111